Mohsin Kadhim Hasan Al Murshdi
Yahya K. M. AL Sultani

The prognostic factors effect on outcome intrauterine insemination

AF670494

MIX
Papier aus verantwortungsvollen Quellen
Paper from responsible sources
FSC® C105338

Printed by Books on Demand GmbH, Norderstedt / Germany

Mohsin Kadhim Hasan Al Murshdi
Yahya K. M. AL Sultani

The prognostic factors effect on outcome intrauterine insemination

The prognostic factors effect IUI

Noor Publishing

Impressum / Imprint
Bibliografische Information der Deutschen Nationalbibliothek: Die Deutsche Nationalbibliothek verzeichnet diese Publikation in der Deutschen Nationalbibliografie; detaillierte bibliografische Daten sind im Internet über http://dnb.d-nb.de abrufbar.

Bibliographic information published by the Deutsche Nationalbibliothek: The Deutsche Nationalbibliothek lists this publication in the Deutsche Nationalbibliografie; detailed bibliographic data are available in the Internet at http://dnb.d-nb.de.

Coverbild / Cover image: www.ingimage.com

Verlag / Publisher:
Noor Publishing
ist ein Imprint der / is a trademark of
ICS Morebooks! Marketing SRL
4, Industriala street, 3100 Balti, Republic of Moldova
Email: info@omniscriptum.com

Herstellung: siehe letzte Seite /
Printed at: see last page
ISBN: 978-3-330-96629-1

Republic of Iraq

Ministry of Higher Education

and Scientific Research

University of Kufa / College of Science

Effect of Spermiogram, Oxidative Stress and Sperm Chromatin Maturity on the Outcome of Intrauterine Insemination of Infertile Patients

Mohsin Kadhim Hasan

Al_Murshdi

بسم الله الرحمن الرحيم

((لِلَّهِ مُلْكُ السَّمَاوَاتِ وَالْأَرْضِ يَخْلُقُ مَا يَشَاءُ يَهَبُ لِمَنْ يَشَاءُ إِنَاثًا
وَيَهَبُ لِمَنْ يَشَاءُ الذُّكُورَ * أَوْ يُزَوِّجُهُمْ ذُكْرَانًا وَإِنَاثًا
وَيَجْعَلُ مَنْ يَشَاءُ عَقِيمًا إِنَّهُ عَلِيمٌ قَدِيرٌ)).

صدق الله العلي العظيم

سورة الشورى/ الآية (٤٩-٥٠)

ACKNOWLEDGMENTS

I thank Allah for giving me the power and health to complete this work.

I wish to express my thanks to my supervisor, Prof, Dr. Yahya K. AL-Sultani for suggesting the title of the study and for his continuous advisement , help and support throughout the study.

I would like to thank the deanery of the College of Science – University of Kufa, Prof. Dr. Merza H. Homady for help and giving opportunity to continue my Ph.D. study.

It is out my duty to express my grateful feelings and Sincere gratitude to the Staff of the Department of Biology in the College of Science, particularly, Prof. Dr. Arshed N. AL-Dujaily, for his kindness, encouraging support and great efforts during the courses of Ph.D. study.

I would like to specially thank Dr. Nezar AL-Nafak who offered great assistance and support by facilitating the difficulties during the nomination to my Ph.D study.

I would like to express my deepest gratitude to Asst. Prof. Dr. Sahib Y. AL-Murshdi for his continuous support and encouragements.

My generous thank are also presented to the staff of the Infertility Center in AL-Sadr Medical City, especially Miss Sanaa and Miss Toah.

My deep thanks are to my friend Assist .Lect Mr. Fares AL-Kufy for helping me and great efforts during the printing contents of this study.

Words of thanks and gratitude won't suffice to reward those who offered great assistance and support by facilitating the difficulties during my research, particularly, my generous family especially my wife and Sons, gratitude and high estimation.

Dedication

To the soul of my Mother

To my Father

To my Family, my wife and my lovely sons

To my brothers

To my sister

To All friends who love me and help me

Summary

The main goal of the current study was to increase the pregnancy rates after IUI by used efficient sperm activation techniques such as density gradient centrifugation and swim up. Also it aims to estimate the impact of other factors on the IUI outcome such as age, infertility duration, type of infertility and smoking.

The study was carried out between January 2014 and October 2014 which including (155) selected infertile men who attended the Fertility Center in Al-Sadr Medical City .

The results showed that significant ($p<0.05$) increase in the semen volume, sperm concentration, progressive motility (A+B) and normal morphology with unexplained infertility compared to other infertile groups. Whereas, the results were recorded a significant ($p<0.05$) decrease in the liquefaction time, leukocyte concentration, MDA concentration and sperm chromatin abnormality with unexplained infertility compared to other infertile groups.

The results also presnted a significant ($p<0.05$) decrease in the sperm concentration after being activaed by both swim-up and density gradient techniques compared to before activation in all infertile groups, while no significant ($p>0.05$) difference between two methods following activation.

There was a significant ($p<0.05$) improvement in progressive motility (A+B) and normal morphology after activation when compared with before activation in all infertile groups, while a significant ($p<0.05$) increase in progressive motility (A+B) via density gradient methods compared to swim-up method was noticed with unexplained infertility, oligozoospermia and asthenozoospermia. Also these results revealed "a significant ($p<0.05$) reduction in the leukocyte concentration, MDA concentration and sperm chromatin abnormality after being activaed by both methods as compared with before activation. A significant ($p<0.05$) reduction was noticed via density gradient compared to swim-up techniques following activation with unexplained infertility and oligozoospermia.

The results in current study also showed a significant ($p<0.01$) increase in the pregnancy rate after IUI with unexplained infertility compared with other infertile groups and a significant ($p<0.01$) increase in pregnancy rate was showed after activation by density gradient method. Moreover, the highest pregnancy rates following IUI were achieved among couples with sperm concentration $>40 \times 10^6$/ml ,progressive motility (A+B) $\geq$ 50% and normal morphology $\geq$ 30% as well as leukocyte concentration $<1 \times 10^6$/ml, MDA concentration < 4 μmol/ L and sperm chromatin abnormality < 30%. The pregnancy rate was a significantly ($p < 0.01$) reduced with age > 40 years, infertility duration > 6 years and with smokers compared to nonsmokers men.

The present study concluded that the defects in sperm parameters and sperm chromatin status as well as excessive levels of oxidative stress contribute to the etiology of infertility of the patients under study. Therefore, semen quality after activation was the most important factor for prediction of successful pregnancy in IUI, then IUI with high quality sperm recovered via density gradient and superior mode of treatment for couples with unexplained infertility .

Contents

List of Tables

List of Figures

List of Abbreviations

A	Asthenozoospermia
AB	Aniline Blue
ANOVA	Analysis of Variance
AOT	Acridine Orange Test
ART	Assisted Reproduction Technology
AT	Asthenoteratozoospermia
ATP	Adenosine Triphosphate
CMA3	ChromomycinA3
COH	Controlled Overian Hyperstimulation
DBD-FSH	Detection Fluorescent in Situ Hybridization Assay
DFI	DNA Fragmentation Index
DNA	Deoxy riboneuclic Acid
Fas	Fatty Acid Synthase
FSH	Follicular Stimulating Hormone
H_2O_2	Hydrogen Peroxide
HCG	Human Chorionic Gonadotropin
HDS	High DNA Stain
HPLC	High- Performance Liquid Chromatography
HSCSA	Human Sperm Chromatin Structure Assay
ICSI	Intracytoplasmic Sperm Injection
IUI	Intrauterine Insemination
IVF	Invito Fertilization

LDL	Low Density Lipoprotein
LH	Leutinizing Hormone
MDA	Malondialdehyde
NT	Nick Translation
O	Oligozoospermia
O2	Super Oxide Anion
OA	Oligoasthenozoospermia
OH	Hydroxil Radicals
8-OhdG	8- Hydroxy-2- deoxy guanosine
OS	Oxidative Stress
P1	Protamine Type One
P2	Protamine Type Two
PBS	Phosphate Buffer Saline
ROS	Reactive Oxygen Species
SCD	Sperm Chromatin Dispersion
SCSA	Sperm Chromatin Structure Assay
SHV	Semen Hyperviscosity
SPSS	Statistical Package for Social Science.
T	Teratozoospermia
TB	Toluidine Blue
TBA	Thiobarbituric Acid
TCA	Trichloroacetic Acid
TUNEL	Terminal Deoxynucleotidly Transferase Mediated Deoxy-Uridinetriphosphate- Nick End Lebeling

UMI	Unexplained Male Infertility
UXI	Unexplained Infertility
WBC	White Blood Cell
WHO	World health Organization
ZP	Zona Pellucida

Introduction

Generally, it is estimated that one in seven couples has problems in conceiving (Skakkebaek *et al.*, 2006). Approximately 15%of couples are infertile patients, 30% associated with male factors, 30-40% by female factors , 20% by male and female together and unexplained infertility associated with 15%.(Filiponi and Feil, 2009). A reduction in the male fertility may be because several conditions like infection of accessory glands, disturbance of endocrine, varicocele, exposure to gonado toxic factors and cancer systemic disease (Hamada *et al.*,2011).

The causes of infertility are not clear in more cases, but some cases are associated with abnormal production of sperm (Peivandi *et al.*, 2010). Oxidative stress (OS) is an important factor which influences fertility potential of spermatozoa by lipid peroxidation which may result in sperm dysfunction (Agarwal *et al.*, 2009). Reactive oxygen species (ROS) with small amount are essential for normal function of sperm, therefore, reactive oxygen species is necessary for the stabilization of mitochondrial capsule in the mid-piece (Desai *et al* ., 2009) The physiological role of ROS is essential in process including: fertilization, capacitation, maturation and acrosomal rection. Furthermore, ROS plays an important pathological role during intrinsic factors such as immotile spermatozoa and leukocytes (Agarwal *et al.*, 2008).

The nuclear chromatin plays an important role for fertilization and embryonic development (Esterhuizen *et al.*, 2000). The proportion of sperm with abnormal chromatin condensations in the ejaculate could be a prognostic factors in assessing the chances of fertilization and pregnancy (Roux *et al.*,

2004). The nuclear compaction is involved in the protection of paternal genome during the transit of spermatozoa through male and female genital tracts and during its interaction with oocyte. Abnormalities in chromatin condensation can cause nuclear damages as DNA denaturation or fragmentation often associated with male infertility (Agarwal and Said, 2003). Many analytical techniques have been proposed to assess nuclear compaction in ejaculated sperm. Some cytochemical or fluorescent dyes were used : aniline blue(AB) staining to evaluate sperm nuclei maturity (Talebi *et al.*, 2008) and toluidine blue staining(TB) used to sperm chromatin damage (Hekmatodoost *et al.*, 2009).

The treatment of infertile patients that obtain in the advance centers by assisted reproduction technology (ART) such as intrauterine insemination (IUI) and intracytoplasmic sperm injection (ICSI) (Turchi and Pescatori, 2005).

Assisted reproduction technologies (ART) are most common as effective in the infertility , in the modern countries services of ART was increased about 5-10% (ASRM,2007). Intrauterine insemination (IUI) is the widely used for the recovery of infertility with different conditions like as unexplained infertility , ovulatory disturbance, cervical mucus hostility and non-severe male factors (Alborzi *et al.*, 2007). Pregnancy rate of IUI varies between researchers, depending on factors of infertility which may include : activation techniques , number of IUI cycles, maternal age and duration of infertility (Badawy *et al.*, 2009). IUI can be used with ovarian stimulation, which increase the number of conception , and increase pregnancy rates than IUI alone (Sikandar *et al.*, 2005). IUI is easy to perform, inexpensive and offers particular advantage such as the minimal equipment required, an easy technique to learn, being less invasive with a reduced psychological burden on the couple when compared to IVF/ICSI (Ombelet *et al.*, 2008).

Sperm activation techniques are essential to sperm activation for ART that involve mainly spermatozoa of normal morphology and progressive motility. Human ejaculation contain various mixture of spermatozoa may which be interactive with leukocyte , debris and other cells (Boomsma *et al.*, 2007). Sperm activation techniques including : density gradient, swim-up and ficoll columns for intrauterine insemination, therefore it provide excellent progressive motility and normal form of spermatozoa, also it removes bacteria, dead sperm and leukocytes produce ROS negatively affecting the fertilization (Henkel and Schill, 2003).

Objectives of the study

1. Evaluation of sperm parameters, malondialdehyde (MDA) levels and sperm chromatin abnormality percent before and after activation for different infertile groups.
2. Determination the effect of sperm parameters, MDA levels and the percentage of sperm chromatin abnormality on the pregnancy rates after IUI of infertile men .
3. Evaluation the efficiency of swim-up and density gradient techniques on the yield of motile spermatozoa in the different infertile groups which lead to increase the rate of IUI outcome.
4. To investigate the relationships between IUI outcome and some of the patients demographic data and medical history (type of infertility, age, duration of infertility and smoking habits).

Literature Review

2.1Male infertility

Infertility is defined as the inability of male to conceive pregnancy in female following one year of unprotected intercourse (Dunson *et al.*,2004). Primary infertility is a term to designate those couples who have never conceived, whereas secondary infertility indicate that at least one conception has occurred but that couple is currently not able to a chive pregnancy (Sharlip.,*et al* 2002). Concerning, the male factors they comprise 50% of all infertility conditions (WHO,2010). Overall factors that lead to infertility they consist of 40-55% in female, 30 -45% in male and approximately 10% in male and female together. Unexplained comprise about 10% (Olayemi, 2010).Most prevalent factors that cause male infertility such as immunological conditions, genital infection and endocrine abnormality (Ferlin *et al* ., 2007). On the other hand, the complete absence of sperm in seminal plasma may leads to male infertility. While, the reduction in sperm count, motility and morphology with abnormal fertilization may cause male subfertility (Francvilla *et al.*,2009). Many environmental, physiological and genetics factors are responsible for defective sperm function which they commonly lead to infertility (Eskenazi *et al.*, 2003). In general, preventing infertility is not easy and does not assist the couple looking for therapeutic recommendation for infertility. Moreover, who are ART is time consuming costly and success rate is low and the failure the rate is high (Bamgbopa, 2010;Lorusso *et al.*,2008). Different percentage of infertility in modern countries approximately of 8-32% of couples (Stephen and Chandra, 1998).

2.1.1 Types of male infertility

Male infertility is explored by routine semen analysis. Routine analysis have been implemented for standard semen parameters including: sperm concentration, motility and morphology. These parameters are given potential information for sperm quality and capacity of fertility (WHO,2010). Male infertility is widely depend on the evaluation of essential sperm characteristics (Guzick *et al.*, 2001). Abnormal production and maturation of spermatozoa are the most prevalent factors of infertility. Sperm could be immotile, abnormal form and immature (ASRM, 2010).

2.1.1.1 Unexplained Infertility

Unexplained infertility is the obscure origin. This case of infertility is obtained spontaneously or because of unknown cause. (Dohle *et al* ., 2010). Unexplained male infertility (UMI) is the category that is concerned for infertile men with normal ejaculation but unknown cause and also without female infertility causes (Sigman *et al.*,2009).After possibility it should be considered the presence of sperm dysfunction and antisperm antibodies with normal semen analysis , should be considered the presence of female factors such as erectile dysfunction and abnormal coital habits. All these Factors lead to unexplained infertility (Honig *et al.*, 1994). The percentage of sperm antibodies which is more than 42% of couples with unexplained and confirm infertility (Moghissi *et al.*, 1980). The treatment with couples with idiopathic male infertility should confirmed the presence of increased values of antisperm antibodies which could be obtained even in the case of normal semen analysis (Francavilla *et al.*, 2007). For men with idiopathic infertility, genetic factors generally play an important role during meiotic division and control spermiogenesis, capacitation and fertilization. Genes expression has become now possible to monitor in the same time with DNA microarray analysis (Garrido *et al.*,2009). Defect sperm may occur with normospermic infertile men which may have an effect on fertilization . This finding depends on the monitoring of decline success percentage of IUI in potential conditions of unexplained patients (Liu and Baker, 2000). The dealing with men with idiopathic infertility may not given typical benefits versus calculation of risks. Therefore, it requires a specific scientific design to diagnosis and correct a known damage (Jarow, 2007). Pregnancy could obtain spontaneously ruled out any treatment in conditions of idiopathic infertility, this advice is for young couples with short duration of infertility(Brandes *et al.*,2011).

2.1.1.2 Oligozoospermia

Oligospermia refers to sperm densities of less than 20 million sperm per milliliter or total count of less than 50 million sperm(Cavallini, 2006). In sever oligozoospermia, hormones studies specifically LH, FSH and testosterone level should be evaluated. If these level are abnormal, a complete hormonal evaluated should be obtained (Sigman and Howards, 1998). Oligozoospermia is classified into three main groups: mild oligozoospermia , the sperm count in this group ranges from 10 to 20 million/ml, moderate oligozoospermia. In this group the sperm count ranges from 5-10 million/ml, and sever oligozoospermia where the sperm count that ranges from 1-5 million/ml (Fauser *et al.*,1990). The health of the sperm in the seminal plasma and viability in the female genital tract as well as several factors such as motility, PH of vaginal fluid, genetic causes and others may lead to oligozoospermia and becoming an important case of infertility (Katosoff *et al.*,1999).

Oligoasthenoteratozoospermia is a condition that reflects lowness in combination of three sperm parametrs (count, motility and morphology) (Isidori *et al.*, 2005). There is no surgical intervention agreed to be efficient (Pauli *et al.*,2008). Oligozoospermia treatment may be very efficient if factors are known but when causes do not identifiy the approach of treatment for oligozoospermia concentration goes on increasing those causes, which stimulate sperm production(Mosammate *et al.*,2009). Approximately 90% of conditions of decrease sperm concentration may because of the low sperm formation. The causes of this condition are unknown and not identified, these cases are labeled idiopathic oligozoospermia (Swerdloff and wang, 2002).

2.1.1.3 Asthenozoospermia

Asthenozoospermia means the reduction of sperm motility or progressive motility or both and is considered one of the most important factors related to male infertility (Dohle *et al.*,2007). Asthenozoospermia may result from a defect in the sperm structure presence of antisperm antibodies, genital tract infection, idiopathic causes, prolonged abstinence period and partial ductal obstruction (Bradsbaw *et al.*,1999). Sometimes, during therapy which may result in decreased motility and this can be improved by the removal of the drug, or introduction of specific culture media to the sperm washing procedure (Ng *et al.*,1990). About 10% of antisperm antibodies are present locally in seminal fluid of infertile couples. These antisperm antibodies decrease the ability of fertilization by negative effect on sperm motility and low cervical mucouse penetration and in vitro interaction of gamete (Lambardo *et al.*,2004). Few studies show no relation between hormonal cause and autoimmunity, although autoimmunity and hormonal imbalance could affect each other to decrease sperm motility because two systems make strong correlation (Bozhedmov and Teodorovich,2005).

Spurious asthenozoospermia is a condition of infertility that occurs by the exposure of sperm to extreme temperature, spermicides, rubber and long delays of examination(Francavilla *et al.*,2009). Semen hyperviscosity (SHV) gives rise to asthenozoospermia through biochemical factors. Past studies have shown that zinc concentration has a positive correlation with sperm motility (Chia *et al.*,2000). Elzanaty *et al* (2004) found that the hyperviscous seminal fluid has been shown to have a negative impact of sperm motility and semen quality. Defect of sperm may be assessed by electron microscope when specific ultrastructural defects are zero sperm motility or less than 5% motile sperm (extreme asthenozoospermia) (Zariwala *et al.*,2007).

2.1.1.4 Azoospermia

A complete absence of sperm from seminal plasma is referred to a zoospermia, two semen samples at least separate centrifuged(AUA,2010). The most common of azoospermia is about 1% among all men while 10% to 15% among infertile men(ASRM and AUA,2001). The initial examination of azoospermia is formed when no spermatozoa may be labeled on high powered microscopic examination of pellet after centrifugation of seminal plasma on at least two separate samples(WHO,1999). Assessment of patients with azoospermia is attended determining the factor and any recovery options that could be divided into three main categories: pretesticular, testicular and post-testicular (Jarvi *et al.*, 2010).

The classification of azoospermia includes obstruction azoospermia which occurred by the obstruction of pathway of the ejaculatory and non-obstruction azoospermia occurred by failed of sperm formation based on the causes such as testicular, post-testicular and pre-testicular (Schlegel, 2004). In azoospermia patients, the initial assessment should consist of physical examination, evaluation of hormones and medical history(Jarvi *et al.*,2010). Difficult collection reasons transit azoospermia or illness. This may be obtained for idiopathic causes with some obstruction and sever spermatogenic disturbance. Spermatozoa could be found in semen which is not completely, the live spermatozoa that may be present. These may be cryopserved for ICSI (Olesen *et al.*,2007). Major causes such as ejaculatory duct pathology, vassal and epididymal may yield obstruction azoospermia. Vassal obstruction may result from vasectomy (Dohle *et al.*,2002), Three main categories pretesticular , testicular and posttesticular azoospermia may result because of genetic factors, the frequency of chromosomal abnormalities is inversely in agreement with the semen concentration, therefore men with azoospermia damage maybe about 10%-15% abnormal. It's on karyotyping in contrast less than 1% in the fertile men(Poongthai *et al.*, 2009). Post-testicular azoospermia may be associated with mutations especially in the cystic fibrosis that lead to regulation gene generally related with congenital vasdeferense abnormalities (Paduch *et al.*,2009).

2.1.1.5 Teratozoospermia

The normal sperm has an oval head and checked on a regular shape with a terminal acrosomal cap which covers the bulk of head, while the mid-piece related to the head is cylindrical in shape and with regular boundaries. The tail shall be cylindrical and non-clad and regular boundaries (Franken and Kruger, 2004). Teratozoospermia (production of >60% morphologically abnormal sperm/ejaculate) is a decrease range of spermatozoa with normal form evaluated by microscope, that is relatively common among various mammalian species(Budhan *et al.*,2006). Strict method is the measurement of the range of sperm having an ideal form. These methods have been added recently in the last edition of WHO for semen analysis (Jacques Auger, 2010). The sperm morphology is the reflection of the development of sperm in the tactical or sperm formation males with damage in sperm maturation which lead to problems with shape of the sperm and may cause failure of fertilization (Alaa *et al.*,2003). Most prevalent sperm with abnormal heads and mid-pieces were referred to sperm defects (Tasdemir and Tasdemir ,1997).Defective of sperm heads vary in sperm head dimensions among and within ejaculating (Garrett and Baker, 1995). In human spermatozoa tail abnormalities range from coiled short or bent tails (light microscope level) (WHO , 2010) and the percentage of cells with such defects are approximately 14% and 50% in fertile and infertile men respectively(Kubo-Irie *et al.*, 2004). Only less than 25% of the motile sperm for fertile males are able to bind the zona pellucida in vitro, The capacity of zona binding is based on to the form of sperm head (Liu et al.,2003).

Many defects including DNA damage, deficiency of centriole and abnormality of chromosome may tend to be abnormal sperm morphology (Chohen-Bacrie *et al.*, 2009). The relevance of the assement of sperm morphology of overall semen sample as it patients to IVF with intracytoplasmic sperm injection(ICSI) is even less clear, oocyte fertilization the though ICSI circumvents many problems, such as poor motility, poor zona binding and poor acrosome reaction which are frequently found in association with specimens having abnormal sperm morphology (Keegan *et al.*,2007).

2.1.1.6 Leukocytospermia

In relation to leukocytes, WHO guidelines indicate that leukocyte is generally present in semen specimen excessively one million/ml is pathologic (WHO,2010). Also an association is found between bacteriospermia and leukocytospermia which stays controversial, especially in sings of males (Lackner *et al.*,2006). Both fertile and infertile males have leukocytes in the semen, even without sex glands infections. As well as, increase leukocyte can be found as feature of epididymis, prostate and seminal vesical. Although, it has been present in 10-20-% of infertile males including for fecundity assessment in the absence a symptoms of sex glands infection (Wolff, 1995).

WHO(2003) indicates that sex glands infection appear in two defective ejaculate symptoms, such as bacteriospermia and leukocytospermia or with sexually disease, tract infection, signs of abnormal physical and epididymitis. (Tomlinson *et al.*, 1992). Leukocytospermia with low levels is also harmful to sperm function and fecundity (Thomas *et al.*,1997). Low levels of seminal leukocyte has been correlated with increased ROS production and elevated sperm DNA damage (Henkel *et al.*, 2005). Recent study have proposed that low-level leukocytospermia is less than the WHO empirical level which could cause sperm defect and decreasing cut off value for the evaluation of leukocytospermia to 0.2 x 10^6 ml (Punab *et al.*, 2003). Although, the present leukocytospermia in semen of infertile patients, used antibiotic therapy which may be a different successful result, because of a decrease in the leukocyte concentration and ROS production following antibiotic therapy and enhanced pregnancy rate (Menkveld,2004). Finally, leukocytes is essential to cause a systemic infection may impact sperm through a condition of influenza, this enhance sperm DNA damages through spermatogenesis (Evenson *et al.*,2000).

2.2 Assisted reproduction techniques

Since 1981, the united states has assisted reproduction techniques to help women to become pregnant. Treatment of fecundity, involving assisted reproduction techniques are commonly successful outcomes including women undergoing assisted reproduction technology and live birth from these procedures (Van Voorhis, 2006). The recovery of infertile patients has made great continuation in the last years through several sperm activation techniques (Boomsma *et al.*, 2007). The treatment by assisted reproduction techniques has become the recovery choice in several conditions of infertility. The semen quality is important factor to determine ART outcomes with present many sperm activation methods are used to isolate spermatozoa (Said *et al.*,2005). Advances in the treatment of infertility using ART have greatly improved the chances of pregnancy (Cantineau *et al.*, 2003). As well as, 8% of children in some western countries are born as a result of ART (Johansson *et al.*, 2011). As the field of assisted reproductively has advanced, many previously untreatedable men are now biological fathers,by the possible use of testicular or epididymal for oocytes fertilization (Katzorke and Kolodziej, 2010).

Quality control is essential for maintenance and improvement of successful ART technique. ART includes intracytoplasmic sperm injection (ICSI), invitrofertilization (IVF) and intrauterine insemination (IUI). (Nyboe and Erb,2006). Advanced assisted reproductive techniques are correlated with enhanced birth disorders which search for other options that may depend on the facility of hospital ,the benefit of the couples and hospital staff (Kumar et al., 2009). In modern countries, services of assisted reproductive techniques have increased to become about 5% to 10%. Enhancing requirement for assisted reproductive techniques are related to increased fecundity in developed populations, which can be associated with delaying to achieve pregnant (Aboulghar, 2005).

2.2.1 Intrauterine insemination (IUI)

Intrauterine insemination with homologous semen suggested as a first selection recovery compared to others assisted reproductive techniques such as ICSI and IVF, in the conditions of cervical, male factor infertility and unexplained infertility. Artificial insemination is used to increase total number of motile sperms at the place of fertilization (Ombelet, 2013).

Contrary to IVF/ICSI methods, IUI is represented a first selection method because its low cost, without essentially risk complication, easy management and simplicity. Although, many causes impacting the probability of pregnancy following intrauterine insemination, such as female factors, semen function, age and body mass index (Souter *et al.*, 2011). Many variables may influence the rate of success after IUI treatment procedures. It is obvious that some factors will never be solved by methodological adjustment involving follicle number, age endometrial thickness and duration of infertility. However, quite a number of variable can be improved upon using different strategies(Macizo *et al.*, 2012).

Although, artificial insemination with men semen stay a widely used treatment selection for many patients with male subfertility (mild and moderate), sexual dysfunction (physiological or psychological), unexplained infertility and cervical factors infertility (Bagis *et al.*,2010).

Regarding selected cervical factors, refered to negative post-coital test, although normal semen and adequate timing, intrauterine insemination in normal cycles significantly increase the probability of pregnancy (Steures *et al.*, 2007). A combination of controlled ovarian hyperstimulation (COH) with intrauterine remains an important option of higher significantly rates of live birth in patients with unexplained infertility (Verhulst *et al.*,2004). It is well known that oocytes and spermatozoa have a limited period of survival. Therefore adequate timing of the insemination seems to be essential . It was generally accepted that insemination should be performed just before or maximally 10 h after ovulation. According to the current available evidence a wider time frame can be followed from 12 to 36h after Human chriong gonadotropin(HCG) injection (Cantineau et al.,2009).

The positive association between pregnancy and the number of preovulatory follicles (>16mm) on HCG day is in accordance with that reported in other studies and it indicates the follicles number which is excellent prognostic predicator of pregnancy rates following IUI (Iberico *et al.*,2004).Most indication of IUI are immunologic abnormalities, ovulatory disturbance, unexplained infertility and cervical mucus hostility(Duran *et al.*, 2002). A study by Badawy *et al.*,(2009) reported that overall success of IUI differ with the rates of pregnancy ranging from 5-70%, which may be due to differences in recovery groups, technique type, age, number of recovery cycles and etiology of infertility.

2.3 Sperm preparation and activation techniques

Sperm activation methods for assist reproduction techniques such as intrauterine insemination were performed to isolate progressive motile and normal form of spermatozoa, and eliminate debris, bacteria, immature sperm and leukocytes produce oxidative stress that negatively affect fertilization. They produce wide range of motile and normal morphology of spermatozoa which might affect recovery option and IUI outcome (Henkel and Schill, 2003). The objective of sperm activation is to isolate the functional sperm from non-functional of the ejaculate. In the condition of intrauterine insemination, to produce many functional motile sperm following activation is a correlation with total number of motile sperm pre-activation that shows the treatment rate (Uranchimeg *et al.*, 2011). Activation methods that have elevation treatment rates are represented better for IUI. The rates of treatment are important when you compare various sperm activation methods (Demirol and Gurgan., 2007). In ejaculation human spermatozoa are unable of in vivo fertilization and should endergo maturational alter through which they get the possibility to fertilize oocyte long time exposure to ejaculation may diminish the possibility of spermatozoa during acrosome reaction in vitro and inhibit fertilization capacity (David, 2000). The major explantation for these results is the activation of motile sperm and normal morphology. The ejaculatory human consists of heterogenouse mixture of sperm with variable motility, leukocytes, epithelial cells, germinal cells and erythrocytes. It is essential to activate sperm specimen for IUI that has normal sperm conformation and progressive motility (Cantineau *et al.*, 2009). The most common sperm activation techniques including: density gradient centrifugation, glass wool columns and swim-up (Agarwal *et al.*, 2006).

2.3.1 Types of in vitro sperm activation techniques

2.3.1.1 Swim-up technique:

Sperm activation techniques such as swim-up and density gradient centrifugation remain the most prevalent techniques, used for the selection of normal and motile spermatozoa. The swim-up remains the essential method for normospermic couples and women infertility. This swim-up is depends on progressive motility of spermatozoa from pre-activation in to overlaying medium (Henkel and Schill, 2003). Swim-up provides normal motile sperm for suspension with decline leukocyte concentration compared to that occurred following density gradient centrifugation (Perticarari *et al.*, 2007). This technique is used for asthenozoospermic and oligozoospermic specimens, when efficient treatment of motile sperm is confirmed through the examination stage (Al-Inany *et al.*,2010; Zavos *et al.*, 2000).Swim-up method is a first line in the sperm activation before assisting fertilization (Oehninger *et al.*, 1990). Several modifications, such as sperm movement into medium from the pellet, a soft or loosened sperm, a concentrated sperm suspension or an aliquot of the ejaculate of the swim-up are described (Carreras *et al.*,1990). According to WHO (1992), the swim-up from semen is considred as the least complex technique for improve the number of motile spermatozoa and it has been recommended for sperm fertilizing ability. The swim-up advantages include the treatment of high rate of motile sperm and without debris and other cells. This technique also yield populations of spermatozoa highly mobile and morphologically normal (Oehninger *et al.*, 1990). Swim-up disadvantage is the decline total treatment of motile spermatozoa, motile spermatozoa down at the bottom of the pellet cannot able to reach and interaction with the culture media. The effectiveness of this method is depended on the original sperm motility in the seminal plasma, also on the compaction levels volume and exposure of final pellet therefore, it limited to seminal plasma with increase motility and sperm concentration (Bossert and De-Jonge, 2007). Another disadvantage of this method which is yield higher levels of oxidative stress and the sperms in pellet, thus coming into close cell to cell contact with leukocytes and debris.

2.3.1.2 Density gradient centrifugation technique

Different methods, most of which involve isolation of cells from seminal plasma by centrifugation. While centrifugation can defective sperm and increase the number of procedure steps increase of probability contamination and interaction with biological substances (Allegra *et al.*, 2007). Density gradient centrifugation showed a better to swim-up and simple method; a best improvement of progressive motility, normal morphology and decrease DNA damages in activated specimen (Boomsma *et al.*, 2007). Generally, spermatozoa activated by density gradient centrifugation is a better compared to swim-up in separation of motile and normal morphology spermatozoa (Morrell *et al.*,2004). The main sources of ROS production in semen are considered leukocytes and abnormal morphology spermatozoa (Drobnis *et al.*, 1991). Therefore, the results by density gradient provide a clean fraction of high motile spermatozoa. The ejaculated volume is used in density gradient producing a significantly increase in total number of normal and motile sperm or having a good yield which it may be used for couples with different degrees of semen quality (Bossert and De-Jonge,2007). Density gradient method dis advantage is that medium used in this technique is more expensive as wll as , time consuming less than used ICSI or IVF. On the other hands, the ROS significantly decrease (Bossert and De-Jonge,2007). The major objective behind the density gradient approach is that the mixture of particles is to be isolated is site on to the surface of vertical column of liquid.

The density which progressive enhances from top to bottom and then centrifugated (Sakkas *et al.*,2000). Although the particles in suspension are individually denser than the liquid at the top of the gradient, the average density for the specimen declines, only under such cases could the specimen zone could be performed by the top of the density gradient (Moedershim, 2010). These sperm preparation methods are still a standard in sperm preparation for ART. Farthermore, exposure to other potential hazards in the laboratort environment such as suboptimal culture media or culture conditions (Sikka, 2004).

2.4 Oxidative stress in male reproduction

Damage sperm quality is the most common cause of infertility and complicated case to recover. Several factors such as genetic ,environmental and physiological have been tend to the decline sperm quality and infertility (Shazia*et al.*,2013). Oxidative stress is an essential cause which affects fertility which essential for spermatozoa via lipid peroxidation which may yield in abnormal sperm. Sperm motility and sperm concentration are potential markers that ascertain the functional capacity of spermatozoa (Abasalt *et al* ., 2013).

In normal condition, reactive oxygen species and antioxidants stay in balance. When the balance is changed into overproduction of ROS ,oxidative stress is produced infertility by two mechanisms . First ,they defect the sperm membrane , reducing sperm motility and its ability to fuse with the oocyte. Second, ROS may change the DNA of sperm , yielding in the passage of DNA damage(El-Tohamy, 2012) . ROS production is normal physiological event in different organs involving the testis, over production of ROS may be evaluated sperm and being related with male infertility.

The increased levels of ROS by leukocytes and immature spermatozoa and via different type of pollutants have been identified as are generated in many metabolic processes (pathological or physiological) in mitochondria

(Gholinezhad and Hosseinzadeh ,2011). Oxidative stress has been induced in several main disease states involving diabetes ,cardiovascular disease , brain disorders and cancer as well as infertility (Tremellen,2008). Abnormal sperm and leukocytes represent main sources of ROS in seminal plasma with cytoplasmic droplets which have been related with excessive damage of sperm DNA by excessive production of ROS (Fischer *et al* ., 2003) . Several enzymatic antioxidants are presented in the seminal plasma such as catalase ,dismutase ,glutathione system and contain a variety of non enzymatic antioxidants such as ascorbate , urate and pyruvate .antioxidants mechanisms have evolved to maintain cells against ROS (Donnelly *et al* ., 1999).

Evaluated of oxidative stress and sperm DNA damage will steer the clinicians and patients towards the best assisted reproductive technique to use for example ,the use of repeated IUI procedure and IVF (Antinori *et al* ., 2008). The diagnosis of damage occurred via increase free radicals against the sperm lipid membrane or DNA. As oxidative stress is the result of an imbalance between production of ROS and total antioxidant capacity, direct tests reflect the net biological impact between these two opposing factors .

The most commonly used technique of evaluating sperm membrane peroxidation is the determination of MDA levels in semen with the thiobarbituric acid assay (Li *et al* ., 2004) .

2.5 Oxidative stress association with male infertility

Several factors such as genetic, environmental and physiological have been induced in the poor sperm quality and fecundity. Free radicals and reactive oxygen species producing oxidative damage to spermatozoa is one such case which is considered essential for its role in producing deterioration sperm quality and infertility(Desai *et a*l.,2010).Increased levels of reactive oxygen species and decrease production of antioxidant enzymes lead to reduced sperm viability and motility and excessive sperm damage because oxidative stress cause damage to lipid, DNA and protein (James *et al.*,2004). Therefore oxidative stress that causes defects to sperm can be mediated during damaged sperm DNA, decline of sperm motility and impaired plasma membrane by lipid peroxidation (El-Tohamy, 2012). Reduction in swollen coiled sperm percentage may be because direct interaction of reactive oxygen species with polyunsaturated fatty acids in the cell membrane causing a chain of chemical reaction, yielding in the formation of different oxidatively modified products, which is toxic to cells. The spermatozoal membrane contains large amount of polyunsaturated fatty acids, which maintain its fluidity, peroxidation of these fatty acids caused the loss of flexibility of membrane and decline in the possibility to swell and when exposed to hypoosmotic solution its expand covering the tail (Amorim *et al.*, 2009). The effect of heaving metals on acrosome integrity can be attributed to the high lipid peroxidation in epididymis because of the excessive oxidative stress, which changes the stability of plasma membrane that surrounds the acrosome during the impact on its content of lipoproteins and poly unsaturated fatty acids (Zini *et al.*,2000). Oxidative stress has been suggested to be involved in male infertility. Its considered a cause of impaired spermatogenesis, diminished sperm migration, poor sperm morphology and affected sperm function (Allamaneni *et al.*, 2005). Therefore, more intense stress can cause necrosis and most sever oxidative stress may causes death to the cells while moderate oxidation may lead to apoptosis (Valko *et al.*,2005). The degree of ROS related defect obtaining in sperm cells based on duration of their exposure to increase free radicals, nature and amount. Large amounts of polyunsaturated fatty acids found in the sperm plasma membrane and low concentration of enzymes in their cytoplasm afford spermatozoa only limited defense strategies against attack DNA (Agarwal *et al.*,2003).

Overproduction of ROS can lead to deterious effects on a range of sperm parameters including count, morphology, and motility and has associated with male infertility. Defective sperm functions were reported to reduce successful fertilization rates with increased miscarriage even after embryo transfer (Henkel *et al*., 2004).Finally, a study by Alvarez(2003) indicated increased ROS production as an independent marker of male factor infertility, uninterested to the normal or abnormal sperm parameters.

2.6 Reactive oxygen species (ROS)

Reactive Oxygen Species represent abroad category of molecules comprised of both radical and radical derivatives of oxygen, nitrogen and other elements (Agarwal *et al*., 2008). A potential compounds in the cell which including nucleic acid, proteins, sugars and lipids are targets for ROS (Zalata *et al*., 2004). Major reaction oxygen species found in seminal plasma such hydroxyl radicals (OH), super oxide anion (O_2) and hydrogen peroxide (H_2O_2) are agents that belong to the class of free radicals and highly reactive oxidizing (Agarwal and Prabakaran, 2005). There are two main sources for the produce ROS in the semen, the leukocyte (mainly granulocytes) and abnormal immature spermatozoa (Twigg *et al*., 1998). In general, spermatozoa produce ROS when spermiogenesis defects occur, with a positive correlation between immature spermatozoa and ROS production (Said *et al*.,2004). The leukocytospermic semen displays a larger percentage of DNA deterioration with might be due to alterations during the regulation of sperm maturation, neutrophils and macrophages are especially associated with an excessive production of ROS (Agarwal and Prabakaran, 2005). The implication of ROS in the etiopathogenesios of various medical conditions, idiopathic infertility, leukocytospermia and spinal injuary (Aarwal *et al*.,2008). In addition a variety of environmental factors including smoking, pollution and obesity are associated with increased levels of ROS (Aarwal *et al.*,2009). Also ROS may affect the development of embryo and fetus health by the modification of the sperm DNA. Free radicals have potential role in the apoptosis process as well as by the activation of the caspases (Martinez *et al.*, 2009). Reactive oxygen species generate the oxidative stress which decrease fluidity of plasma membrane and fail its function (Saleh and Agarwal, 2002).Moreover, this reduction in fluidity might affect the transport activity of membrane and by affecting on the vitality of sperm. Lipid peroxidation affects the sperm morphology, motility, concentration and associated with defect sperm function (Hsieh *et al.*,2006). ROS generation serves as a defence mechanism to combat invading pathogens. Leukocytes can produce ROS directly as described above, or indirectly through the activation of Pro-inflammatory cytokines, were correlated with high levels of ROS (Agarwal and prabakaran, 2005). The important role of ROS is a second messenger in different types of cells. Also, if their ROS production uncontrolled, it becomes essential factors in the causes of male infertility in particular and pathologic cases in general (Aitken and Fisher, 1994). Nevertheless, they have pathological impacts on macromolecules such sugars , amino acids and polyunsaturated fatty acids in high levels (Agarwel *et al.*, 2003). The main pathological effect of ROS that lipid peroxidation is related with oxidation of membrane polyunsaturated fatty acid (Fraczck *et al.*, 2001). The correlation between excessive ROS production and deterioration of sperm function show in the found of increase cytoplasmic droplet (residual cytoplasmic) when spermatogenesis failed, the cytoplasmic extrusion strategies get damage and spermatozoa are released from the germinal epithelium carrying excessive cytoplasmic droplet, the spermatozoa that are released through spermiation is believed to be immature and causes damage (Huszare *et al.*,1997).

2.7 Sperm DNA abnormality and male infertility

The nuclear DNA is sited in the head of sperm and generally named the genome. Mitochondrial DNA represent the second type and is responsible for delivering the spermatozoa to the ova by providing energy for cells acceleration. Two types of DNA play and important role for fertilization process (Lewis and Aitken, 2005). Sperm and ova carry all the necessary information required for fertilization and embryonic development. Any form of damage to the paternal (sperm DNA) or maternal (ova) can have serious consequences in the form of pre and post implantation losses, impaired embryonic development, childhood morbidity and even cancer (Kumar *et al.*,2012). The major forms of DNA damage in male sperm involve mutation, DNA fragmentation, base oxidation, epigenetic modifications and chromosomal aberrations (Salah *et al.*,2002). Aitken *et al.*,(2010) suggested the sperm DNA which may pose problems in the process and sperm DNA also occur during spermatogenesis toward the apoptotic process or through the compaction of sperm chromatin and also as histones are replaced by protamines. In such instances, these ROS attack DNA bases and phosphodiester backbones causing DNA fragmentation (Villegas *et al.,*2005). Sperm DNA damage subsequent impacts of such defect on male fertility, is the main factor which has been studies in the last years. Because half of the offspring DNA is yielding from paternal unit of sperm chromatin and DNA damage on reproduction (Saleh *et al* .,2002).

There is a need to develop new markers that differentiated from fertile to infertile men and help to predict ART outcome and contrary reproductive process. Tests of sperm chromatin and DNA integrity can assist in the assessment of male infertility (Zini *et al.*, 2001). Moreover, a study by Bungum *et al* (2007) reported that tests of DNA damage and sperm chromatin may help in predicting reproductive outcomes after using reproductive technologies but the value of these tests in the clinic remains to be fully delineated. The breaks of sperm DNA strand are referring to deteriorated quality. Damaged sperm DNA can obtain because the following strategies: first, deficiency of protamine, a genetic disturbance that is most prevalent infertile couples, in which the DNA protective function is reduced (Carrell and Liu, 2001). Second, sperm DNA damage also correlated with oxidative stress induced by varicocele and leukocytosis (Salah *et al.*,2003) and third, exceed activity of apoptotic in older couples and in those with defect semen function can also attribute to DNA damage (Seli *et al.*,2004).

2.8 Etiology of sperm DNA damage

Spermatogenesis is the process by which the testes produce spermatozoa by meiosis . This process occurs in all semniferous tubules during active sexual life(de Kretser *et al* .,1998).Where sperm DNA damage or its defective of sperm chromatin may obtain at any step(Erenpreiss *etal* .,2006).It has been suggested that DNA defect in sperm may be due to unrepaired DNA breaks through the spermatogenetic chromatin remodeling and packaging or abortive apoptosis through spermatogenesis . Other etiology is the impact of endogenous endonucleases and caspases, exposure to several genotoxic agents due to therapeutic causes or because of occupational or exposures of environmental factors and lastly , the effect of oxidative damage (Sakkas and Alvarez, 2010). The crossing over events represented a defective checkpoint through spermatogenesis or protamine deficiency will lead to the sperm becoming more defecting by oxidative stress, the imbalance between the antioxidant in the ejaculation and ROS production cause the formation of oxidative stress which leads to impaired plasma membrane by formation of lipid peroxidation as well as attack sperm DNA (Aitken and krausz,2001). Sperm loss DNA repair system and antioxidant (Aitken *et al* .,2003), and maintenance of the offspring from the negative impacts of DNA strand beaks resulting from the spermatozoa overall base on the repair possibility of the oocyte and the early embryo . Immature spermatozoa and leukocytes in seminal plasma are the major sources of ROS in semen (Sakkas and Alvarez, 2010). Several cause of sperm DNA damage such as smoking , sex gland infection ,chemotherapy and varicocele has been correlated with damage of sperm DNA and a abnormal chromatin structure (O'Flaherty *et al.,* 2008).Other studies have proposed that deficiency of protamin , reactive oxygen species , abortive apoptosis and alterations in topoisomerase II activity can lead to the damage of sperm DNA (Leduc *et al.*,2008; Tarozzi *et al* ., 2007).

To explain the originating of sperm DNA damage , two-step hypothesis depend on oxidative stress act on decline cells protaminated which in complete histones replacement by protamines , generated as a result of defective spermatogenesis (De Luliis *et al.*, 2009).A host of pollutants and environmental exposure have been related with alters in sperm quality (Loft *et al* ., 2003). More of these factors cannot only disorder hormone level but also yield oxidative stress, which sperm DNA might get damage (Bennetts *et al* ., 2008). Furthermore, the correlation between cigarette smoking and DNA damage has been showed and smoking has been established to affect the integrity of sperm DNA .Therefore ,smoking is correlated with poor semen function (Kunzle *et al* ., 2003).

2.9 The Main mechanism of chromatin abnormality

The exact hypothesis by which DNA damage/ chromatin abnormalities generated in human spermatozoa is not precisely understood (Agarwal and Said,2009).Three major hypothesis have been suggested: oxidative stress ,apoptosis and defective sperm chromatin packaging .

2.9.1 Oxidative stress

Reactive oxygen species play an essential role by differentiating, functioning and modulating cell, but it can have negative impacts too because these species are highly reactive and may defect any cell structure involving the DNA molecule (Aitken *et al.*, 2010) .In the ejaculation of fertile men the amount of ROS formation is exactly controlled by seminal antioxidants . The pathogenic impacts of ROS obtained when they yield in increasing of the antioxidants capacities of the male reproductive tract or seminal plasma . (Aitken and curry ,2011).Also ROS may be causes chromosome deletions ,break of DNA strands , chromatin packaging ,chromatid exchanges and dicentrics (Agarwal and said, 2003) ROS induce oxidative stress which has been associated with high frequencies of single and double DNA strand breaks and apoptosis (Aitken and Krausz,2001).

2.9.2 Apoptosis

Apoptosis is a main events obtaining through spermatogenesis that controls the excessive production of male gametes to numbers that may depend on the sertoli cells and option of sperm cells with undamaged DNA (Smit ,2011). Sertoil cells may start and order germ cell apoptosis during the apoptosis stimulating fragment system that is characterized by the reaction between fatty acid synthase (Fas) and Fas ligand . Cells to be eliminated will enter a system cascade of reaction leading to apoptotic cell death (Ozmen *et al.*, 2007). Fas cell surface proteins assist regulation of the apoptosis in spermatozoa and correlated ligand may serve as markers for the apoptosis rate. In males with abnormal semen function , the percentage of Fas-positive spermatozoa may be as high as 50% (Francavilla *et al* ., 2000). Therefore, these Fas positive cells that release apoptosis , can successfully mature and their damaged DNA may be included into the gene pool (Singh and Agarwal, 2011).The abortive apoptosis obtained in different couples which show the decline of sperm quality. In certain couples, abortive apoptosis may fail in the total elimination of spermatozoa limited for removal by apoptosis .Therefore, the contamination of ejaculated spermatozoa found an array of anomalies that are representative of the characteristics noticed in cells undergoing apoptosis (Wang *et al* .,2003). intracellular ROS produced by mitochondria are thought to regulate apoptosis . This is supported by the observation that sperm from infertility patients often have significantly higher rates of both apoptosis and ROS than donor sperm (Hughes *et al.*, 2005).while other investigators have presented no association between DNA damage and apoptosis , it hers have demonstrated increased levels of apoptosis in poor quality spermatozoa (Barroso *et al* .,2000).

2.9.3 Defective chromatin packaging

Mammalian sperm DNA is highly organized into linear groups regulated as loop domains . In comparison with other types ,human sperm chromatin packaging is exceptionally deferent both within and between individuals . This variety is widely contributed to its main protein (Aitken and Deluliis.,2007) .15%of histones ,which are less basic than protamines, making the generation of less-compact chromatin structure (Aitken and DeLuliis,2010). Furthermore, contrary to the ram, boar, cat and bull, whose spermatozoa include only one type of protamine (p1), human spermatozoa involve another type of protamine called protamine (p2).Changed p1/p2 ratios and without(P2) are related with problems of male fertility(De Mateo *et al* ., 2009) . Abnormal spermatozoa show high damaged of DNA and production of ROS and possibility to have change in predomination and chromatin packaging. Fragmentation of sperm DNA is described by single and double strand DNA breaks which are usually observed in the ejaculates of infertile males(Rousseaux and Prevoust ,1995). Our study has appear that oocyte and early embryos may repair damaged of sperm DNA. In addition

,the biological impact of defect sperm chromatin structure based on the involved impacts of sperm chromatin defect and the capability of the oocyte to repair it. Any errors that may obtain through this post-fertilization period of DNA repair have the essential to create mutations that can effect development of fetal and finally, the health of the child (Cocuzza *et al*., 2007). Chromatin packaging involves a step that require endogenous nuclease effectiveness to separate the chromatin by histone hyper-acetylation and introduction of breaks by topoisomerase II, capacity of both creating and ligating breaks chromatin packaging around the new protamine cores should be completed and integrity of DNA sited through epididymal transit (Mengual *et al* ., 2003) .

2.10 Human sperm chromatin structure assays (HSCSA)

Different assays have been used to assess DNA integrity and sperm chromatin (Chan *et al* .,2001). The sperm chromatin structure assay (SCSA)is used to detect the sperm chromatin structure abnormality using high DNA satiability (% HDS) and DNA fragmentation index (%DFI) due to abnormal nuclear proteins . These assays depend on the premise that sperm DNA with defective chromatin structure is more exposures to heat denaturation or acid(Schulte *et al* ., 2010). According to approaches, their two way which one of them the methods only detect breakage by using methods such Comet (sperm chromatin dispersion SCD).This method has been categorized as a simple and inexpensive technique for the analysis of sperm DNA fragmentation (Fernandez *et al* ., 2003). Many methods have been used to the assessments sperm chromatin abnormality and DNA integrity (table 2-1).These methods involve simple staining methods such as toluidine blue (TB) and aniline blue (AB) stains ,fluorescent staining methods such as DNA breakage detection fluorescent in situ hybridization assay (DBD-FSH), chromomycin A3(CMA3),sperm chromatin dispersion (SCD), cytometric based sperm chromatin structure assay and in situ nick translation (NT) (Evenson and Wixon ,2006).

Although two methods appear to be associated with result ,they are not equal and appear several types of defect ,especially the terminal deoxynucleotidly transferase mediated deoxy-uridinetriphosphate-nick end lebeling (TUNEL) assay . the method uses TdT to label the $\bar{3}$- OH (Evenson *et al* ., 2007) .While indirect assays such the acridine orange test (AOT) or sperm chromatin structure assay (SCSA)(Bungum,2012). High – performance liquid chromatography (HPLC) which measures the concentration of 8-hydroxy -2- deoxy guanosine (8-OhdG), a by-product of oxidative damage of DNA in the sperm chromatin (Agarwal and Said,2009).

Table(2-1) Various methods for assessing sperm chromatin abnormalities (Agarwal and Said, 2009).

Assay	parameter	Method of analysis
Acidic aniline blue	DNA Maturity (Nuclear maturity & protein composition)	Optical microscopy
Toluidine blue Stain	DNA fragmentation	Optical microscopy
Chromomycin A3	DNA Maturity (Nuclear maturity & protein composition)	Fluorescent microscopy
DNA breakage Detection- Fluorescent	DNA fragmentation (ssDNA)	Fluorescent microscopy
Acridine orange	DNA denaturation (acid)	Fluorescent microscopy Flow cytometry
TUNEL	DNA fragmentation	Optical microscopy Fluorescent microscopy Flow cytometry
Comet(neutral)	DNA fragmentation (dsDNA)	Fluorescent microscopy
Comet (alkaline)	DNA fragmentation (ssDNA/dsDNA)	Fluorescent microscopy
Sperm Chromatin Structure	DNA denaturation (acid/heat)	Flow cytometry
8-OHdG measurement	8-OHdG	High-performance liquid chromatography

2.11 Malondialdehyde (MDA)

The key mechanisms of ROS is lipid peroxidation of sperm membrane which leads to sperm defect and causes infertility (Shilpa and Bijoor ,2013).

Malondialdehyde is one of the most common used markers of lipid peroxidation , a by–product of lipid peroxidation processes (Rumely *et al.*,

2004)Rosenfeld et al (1990) demonstrated that MDA modified low density lipoprotein (LDL)can be recognized and taken up by macrophages via scavenger receptors, causing massive internal accumulation of cholesterol esters. Seminal plasma MDA products, are an easy technique to assess the impact of lipid peroxidation on sperm (Geva *et al.*,1998). Malondialdehyde may be determined by using various methods ,the most prevalent technique is the thiobarbituric acid reactive substances assay (Lee *et al.*, 2012). Although the association or exact role of malondialdehyde with semen function stays controversial (Pasqualotto *et al.*, 2000). Malondialdehyde is a marker for lipid peroxidation and as an end – product of peroxidation lead to the loss integrity of cell membrane , making exceeded in permeability of cells, enzyme activation DNA damage and cell death (Sies,1991).

2.12 Aniline blue (AB)

Sperm head maturity is a potential for detection the capacity of sperm fertilizing. Indeed of assessment the maturity of sperm head with routine semen analysis increase the evolution and option of infertility recovery (Sellami *et al* ., 2013) . Aniline blue staining assessment the compaction degree of sperm chromatin and defect of sperm chromatin. This assay allows discriminating between found of arginine and cysteine rich protamines and lysine rich histones in sperm nuclei (Agarwal and Said ,2009) .Consequently ,immature sperm has been histone rich nuclei will take up the blue stain whereas mature sperm contain protamin rich nuclei remain unstained(Hofman and Hilscher ,1990).

Materials and Methods

3.1: Patients:

The study was carried out in the Fertility Center in Al-Sadr Medical City, through the period from January 2014 to October 2014 .

Semen samples were obtained from one hundred and fifty five (155) infertile men they were divided into six subgroups. (figure 3.1).

According WHO criteria (1999) the semen quality of infertile patients were divided into:

a) Unexplained infertility(UXI).
b) Oligozoospermia (O) .
c) Asthenozoospermia (A).
d) Teratozoospermia (T).
e) Oligoasthenozoopermia (OA).
f) Asthenoteratozoospermia (AT).

According to age, it was divided into:

a) Patients with ages less than 30 years.
b) Patients with ages between 30-34 years.
c) Patients with ages between 35-39 years.
d) Patients with ages $\geq$ 40 years .

According to type infertility, it was divided into:

a) Primary.
b) Secondary.

According to duration of infertility, it was divided into:

a) Patient with duration of infertility between 1-3.
b) Patient with duration of infertility between 4-6 years.
c) Patient with duration of infertility > 6 years.

According to smoking habit, it was divided into:

a) Smokers.

b) Non – smokers .

Demographic and medical history information was taken from the patients before starting the evaluation of semen analysis. This information was recorded in a prepared data form (figure 3.2) .

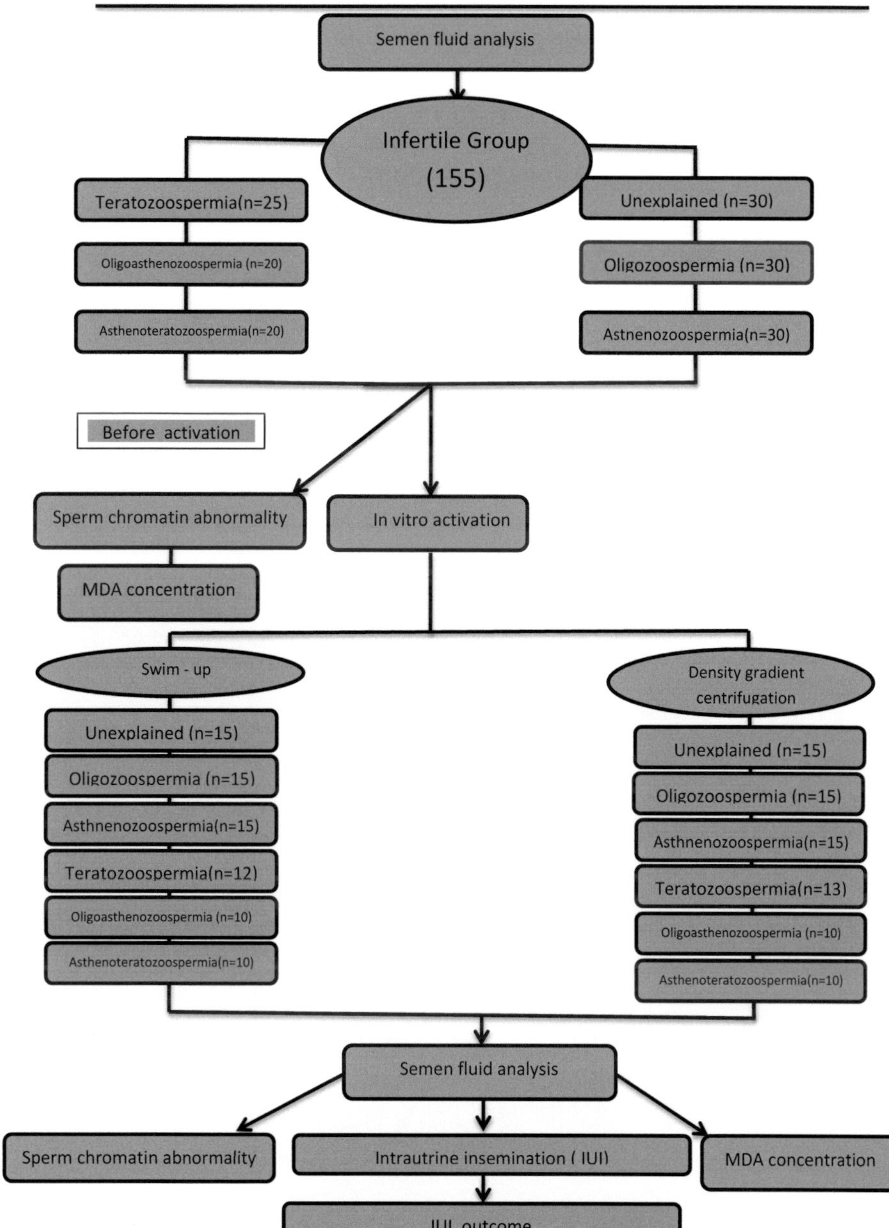

Figure(3.1): The Experimental design of the study

Patient No [ ]	Data: / /
Name:	
Husband Age [ ] years	
Wife Age : [ ] years	
Full address:	
Profession:	
Type of infertility:	
Primary [ ]	secondary [ ]
Infertility Duration [ ] years	
Smoking [ ]	
Abstinence period [ ] days	

Figure (3.2) The form of the medical history for infertile patients

3.2: Materials

3.2.1: Tools and equipment:

Name	Origin
- Centrifuge	Universal 16 A , Germany
- Electric balance	Sartorius , Japan
- Eppindorf tube (10-20μl)	Germany
- Filter (0.22mm)	Millipore, USA
- Incubator	Memmert, Germany
- Light microscope	Olympus, Germany
- Micro – pipette	China
- Spectrophotometer	APEL PD-303 UV, APELCO LTD
- Tips	Gilson, France
- Water bath	Memmert, Germany

3.2.2: Chemicals and Culture media

Name	Origin
Acetone	BDH, England
Aniline blue	BDA, England
Ethanol	Sigma – chemical, USA
Ferticult flushing medium	Fertipro. N.V Belgium
Glacial acetic acid	Sigma – Aldrich, UK
Glutaraldehyde	Sigma – chemical, USA
Phosphate buffer saline (PBS)	BDH, England
Sill – select stock medium	Fertipro. N. V Belgium
Thiobarbituric acid (TBA)	BDH, England
Trichloroacetic acide (TCA)	Thomas Baker (India)

3.3: Seminal fluid analysis

Semen sample should be produced by masturbation after 3-5 days of abstinence, in wide mouth disposable plastic container. Immediately placed in an incubator at 37C° till complete liquefaction . After liquefaction time, the liquefied semen was then carefully mixed for few second, and the analysis of semen samples before and after activation were done. The specimens were examined by macroscopic and microscopic examinations using standardization of WHO (1999). Table (3.1).

Table (3.1): WHO normal values of semen (WHO, 1999)

Macroscopic examination		Normal value
Appearance		Homogenous /opalescent Gray
Volume :		2- 6 ml
Liquefaction time		$\leq$ 60 minute
Viscosity		Drops $<$ 2cm thread
pH		Alkaline (7.2-8.2)
Microscopic examination		Normal value
Sperm concentration		$\geq 20 \times 10^6\ sperm/ml$
Sperm Grade activity (%)	Grade A	Progressive motile sperm (Grades A+B) $\geq$ 50% ; within 60 minutes
	Grade B	
	Grade C	
	Grade D	
Morphology sperm (%)	Normal	$\geq$ 30%
	Abnormal	$\leq$ 70%
Sperm agglutination		$<$10 sperm / agglutinate
Leukocyte count		$< 1.0 \times 10^6\ /ml$

3.4: Macroscopic examination

Semen sample was assessed directly following liquefaction or through one hour of ejaculation at room temperature for recording the parameters including, semen appearance, volume, liquefaction time, viscosity and acidity (pH) that can be studied by naked eyes (figure 3.3) .

3.4.1: Appearance :

Normal semen is a homogenous grey-opalescent color and variations in this color may be indicative of disease .If a sample is oligospermic then it is less opaque .A red/brown color indicates the presence of red blood cells .This could be indicative of infection .Semen can also have a yellow appearance due to jaundice or some vitamin supplements (WHO,1999).

3.4.2: Volume

Normal ejaculate volume is between 2-6 ml (WHO, 1999). The volume was measured by using graduated centrifuge cylinder with a conical base. The semen sample was considered hypovolemic when the volume is less than 2 ml or hypovolemic if the volume is more than 6ml .

3.4.3: Liquefaction

The liquefaction is the process when the gel formed by proteins from the seminal vesicles is broken up and the semen becomes more liquid . It is normal liquefied < 20 min. for the specimen to alter from a thick gel into Liquefied at room temperature. Abnormal liquefaction with long time (> 1 hr at room temperature or 30 minutes at 37C°) may indicate an infection (WHO, 2010). The sample was mixed in the collecting container before microscopic examination. Occasionally , semen sample may not liquefy, probably due to a poor prostate activity or gel particles or mucous streaks (NAFA and ESHRE, 2002), in which case additional treatment was applied by using mechanical mixing with a syringe to gauge needle 20.

Semen Analysis Report				
Patient's name: Days of abstinence: Time of collection: Time of examination: Date:				
Macroscopic examination				
Parameters	Results			Normal values
Appearance				Homogenous, opalescent gray
Volume (ml)				2-6 ml
Liquefaction time (min)				Within 60 min
Viscosity (cm)				< 2 cm
pH				7.2-8.0
Microscopic examination				
	Before activation	After activation		
Sperm concentration ($\times 10^6$ $sperm/ml$)		Swim-up	Density gradient centrifugation	($\geq 20 \times 10^6$ $sperm/ml$
Sperm motility (%)				$\geq$ 50%
Progressive sperm motility (%)				$\geq$ 50 % with forward progressive motility
Normal sperm morphology (%)				$\geq$ 30 %
Leukocyte count ($\times 10^6$ $/ml$)				$< 1.0 \times 10^6$ $/ml$
Other tests				

Figure (3.3) Macroscopic and microscopic examination scheme of the present study

3.4.4: Viscosity

Viscosity of semen specimens was estimated by using pasture pipette. The specimens is considered normal when it is easily drawn up into a pipette as small drop. Abnormal specimens (highly viscous)will form a thread greater than 2cm long(WHO,1999).

3.4.5: pH

Semen pH was measured by immersing graduated litmus paper (ranged from 6 to 14) in the semen sample while it is in the petri – dish . The normal pH of semen is slightly alkaline and showed be equal to or more than 7.2 (WHO,1999) .

3.5: Microscopic examination

For each sample a drop 10µl of liquefied which is thoroughly mixed is delivered on to a microscopic slide and covered with a standard cover slip (22 × 22mm). The sample is spreading for optimum viewing by the weight of the coverslip. Then, the preparations were scored under magnification of 40x objectives. Specimens were examined for the following parameters(fig 3.3).

3.5.1: Sperm concentration

Sperm concentration per milliliter (ml) was reported from the mean number of spermatozoa in 10 random microscopically fields. This number was multiplied by factor of (1×10^6) . Semen sample with concentration of less than 20 million/ml was defined as an oligozoospermic semen sample (WHO,1999).

Sperm concentration = Mean No. of sperm × million

3.5.2: Sperm motility and grade activity

Fifty percent or more of the sperm should be moving, which describe the percentage of sperm that are moving.

One drop (not more than 10 µl) of liquefied thoroughly mixed semen was mounted between a warmed slide and covered with a standard coverslip. Determination of sperm motility was taken under room temperature. The number of motile spermatozoa in the ten randomly selected field was counted a way from coverslip edge. The mean number of progressively motile spermatozoa was calculated.

Sperm grad activity was defined as that with a forward progressive motion, which showed defined space gain or an approximate linear velocity. Non progressive spermatozoa were defined as those motile spermatozoa that did not show defined space gain or linear velocity and included those spermatozoa showing only feeble flagellar beating. Immobile spermatozoa were those which showed no flagellar movement at all.

The percentage of average of (A + B) progressive motility was calculated by using the following equation:

$$progressive\ motility\ (A+B)\% = \frac{No.of\ (A+B) progressive\ motile\ sperms}{Total\ No.of\ sperms} \times 100$$

3.5.3: Sperm morphology

Spermatozoa were examined for normal and abnormal morphology. The normal head should be oval in shape. The morphologically abnormal sperm was noticed in head (primary abnormalities) as macrocephalic, tapered, double, round, amorphous and pin head. Abnormalities were also reported in the sperm tail (secondary abnormalities) as either double, short, coile, bent or broken tails.

Cytoplasmic droplet was seen as around particle attached to the midpiece indicating abnormal sperm maturity in the epididymis or it could be due to frequent ejaculation (WHO, 2010). At least the percentage of morphologically normal sperm was reported from the following formula:

$$Normal\ sperm\ morphology\ (\%) = \frac{No.normal\ sperm\ shape}{Total\ no.\ sperm\ (normal\ and\ abnormal)} \times 100$$

3.5.4: Leukocyte:

To distinguish the W.B.C. from immature sperm, round cells, were assessed with peroxidase staining (Shekarriz *et al.*,1995). Peroxides-positive Leukocytes staining brown were counted under bright-field microscopy and results reported as $\times 10^6$ WBC/ml semen.

Semen infection is considered mild leukocytospermia when leukocytes concentration between (1 and 2/mL), while it is considered marked infection between (2 and 5 /mL). Excessive leukocytes concentration above (5/mL) were considered very marked infection. (Hinting, 1989).

3.6: Sperm activation techniques

Two methods of sperm activation have been used in this study to separate the motile spermatozoa from seminal plasma .(Bjorndahl *et al.*,2010).

a) Swim – up technique .
b) Density gradient centrifugation technique.

3-6-1 Swim-up technique

Including the following steps have been done:-

1-Mix the semen sample well.

2-Dilute the semen sample 1:1with ferticult medium to promote removal of seminal plasma.

3-Transfer the mixture into centrifuge tubes.

4-Cenrifuge at 2500 rpm for 5 minutes , discard the supernatants .

5- Gently, layer 1ml of ferticult medium over the pellet and incubate for 60 min at 37C°.gently, remove the upper part of medium which will contain highly motile sperm cells.

6-Adrop (10μ) was taken and put on a slide and covered with a cover slip then examined at a microscope under 40x objective for assessment of sperm function parameters. Also MDA concentration and sperm chromatin abnormality percent were estimated.

3.6.2: Density gradient centrifugation technique

The density gradient centrifugation selects spermatozoa according to their density and gravity. Including the following steps have been done:

1. A two layer gradient was prepared using respectively solutions of 90% (lower layer) and 45% (upper layer) sil select stock medium.
2. Place all components of the upper and lower layer in incubator for twenty min at 37c° .
3. Transfer 2ml of the lower layer in to bottom of sterile conical centrifuge tube.
4. Layer 2ml, of the upper layer on top of the lower layer using a transfer pipet. Slowly dispense the upper layer lifting the pipet up the side of the tube as the level of the upper layer rises. A distinct line separating the two layers will be observed. This two – layer gradient is stable for up to two hours.
5. By using a sterile pipette, 1ml of semen liquefy was put on to upper layer in a centrifuged tube.
6. Centrifuge at 3000rpm for 10 minutes.
7. Removal supernatant and pellet was mixed with 3ml of ferticult medium.
8. Centrifuged at 3000rpm for 10 minutes.
9. Mixed the final pellet with 0.5- 1 ml of ferticult medium.
10. Then, certain sperm function characters are recorded. Also MDA concentration and sperm chromatin abnormality percent were estimated.

3.7: Determination of malondialdehyde (MDA) concentration

Seminal MDA levels were analyzed according to (Muslih *etal.*, 2002) was assessed using the thiobarbituric acid method.

3.7.1: Preparation of reagents

1) Trichloroacetic acid (TCA) 70%, dissolved 70g of TCA in distilled water and complete the volume into 100ml .
2) Trichloroacetic acid (TCA) 17.5%, dissolved 17.5g of TCA in distilled water and complete the volume into 100ml .
3) Thiobarbituric acid (TBA) 0.6% . 0.6g of TBA were dissolved in a final volume of 100ml distilled water.

3.7.2: Procedure

The test tube contain 0.15ml of semen sample, 1ml of 17.5% TCA, and 1ml of 0.6% TBA. The tubes were vortexes and boiled in a water bath for 15min, then allowed to cool. Added 1 ml of 70 % TCA and left the mixture at room temperature for 20min. Centrifuged for 15min at 2000 rpm and the absorbance of supernatant was read on a spectrophotometer at 532nm.

MDA concentration was calculated according the following equation.

$$MDAcon. (\mu mol/L) = \frac{A532}{L \times E_o} \times D$$

A: Absorbance at 532nm.

L: Light bath = 1cm.

E_o: Extiniction Coefficient = $1.5 \times 10^5 M^{-1} cm^{-1}$.

D: Dilution factor = $1ml/0.15ml = 6.7$.

3.8: Aniline blue staining

3.8.1: Preparation of AB stain

The fixative solution was prepared by adding (3ml) glutaraldehyde to (97ml) phosphate buffer saline PBS, to obtain (3%) concentration of glutaraldehyd in PBS. Staining solution was prepared with (5g) AB per 100ml PBS, then placed in water bath for a short time, then it was filtered and added drops of glacial acetic acid to adjust a pH of (3.5) .(Hammadeh *et al.*, 2001).

3.8.2: Staining technique by (AB) stain

Slides were prepared by smearing 10 µl of semen specimen. The slides were left to dry in air and then fixed with 3%glutaraldehyde for 30 min. The smear in stained with 5%aqueous AB solution (PH3.5)for 5 min. sperm head with immature chromatin condensation will stain blue and those with mature chromatin will not be take the stain (Sellami *et al.*,2013).

3.9: Statistical analysis

It was done by mean of SPSS (Statistical Package for Social Science; Version (20.0) . Results were expressed as mean ± slandered deviation (S.D). To compare between the parameters of two groups, independent sample t- test was used . To compare the parameters of three groups or more ANOVA was used . The differences between values were considered statistically significant at $p < 0.05$ and $p < 0.01$.

4-1: Semen parameters, seminal MDA level and sperm chromatin abnormality percent for infertile patients:

Table (4-1) shows the result of seminal fluid analysis for all infertile groups including: unexplained infertility, oligozoospermia, asthenozoospermia, teratozoospermic, oligoasthenozoopermia and asthenoteratozoospermia.

It was clear that semen volume has a significant (P < 0.05) difference among all the above infertile groups, whereas the results recorded the high value with UXI (3.8±0.59) but the less value of semen volume was showed in AT (2.9±0.89) . Also there was a significant (P < 0.05) differences in the liquefaction time which was found in infertile groups, but the decline value was showed with UXI(26.4±7.05). The higher value in liquefaction time appeared with AT(43.9±6.61).

There was a significant (P < 0.05) difference in the mean of sperm concentration . It was showed among the infertile groups, therefore, increase value was observed with UXI (69.1±6.46) while the reduce value showed both O (16.7±1.51) and OA (15.3±2.55) whereas , no significant (P > 0.05) difference between two groups.

On the other hand, the percentage of progressive motility (A+B) showed a significant (P < 0.05) difference among the infertile groups, the higher value was recorded with UXI (65.1±5.99) while the less value was found with OA (26.7±7.81) and then followed AT (32.3±3.9) and A (38.73±6.18). No significant (P < 0.05) difference was observed between O and T.

According to the percentage of normal morphology, there was a significant (P < 0.05) difference among the infertile groups, while the increase value was observed with UXI (58.5±7.76) but the reduce value was showed with AT (22.4±3.29) and T (24±3.75). Moreover, there was a significant (P < 0.05) difference in the Leukocyte concentration among the infertile groups. The higher value was found with AT (4.37±1.29) then decline value was recoded with UXI (1.66±0.76). No significant ($P > 0.05$) difference appeared between O and A.

Recording the results of MDA concentration showed a significant ($P < 0.05$) difference among the infertile groups . The reduce value was noticed with UXI (4.7±0.44) but the higher value was showed with AT (8.6±1.65).

Lastly, a significant ($P < 0.05$) difference in percentage of sperm chromatin abnormality was recorded among infertile groups. The decline value was showed with UXI(32.1±3.09) while higher value was observed with AT (45±6.18) .

Table(4-1) Semen parameters, Seminal MDA Levels and Sperm chromatin abnormality percent for infertile patients before activation .

Patients / Variables	UXI n=30	O n=30	A n=30	T n=25	OA n=20	AT n=20
Semen volume (ml)	3.8±0.59 ^{a}	3.6±0.54 ^{b}	3.5±0.83 ^{c}	3.3±0.7 ^{d}	3.2±0.62 ^{e}	2.9±0.89 ^{f}
Liquefaction time (min)	26.4±7.05 ^{a}	35.4±3.03 ^{b}	39.1±3.99 ^{c}	37.8±4.41 ^{d}	40.5±4.65 ^{e}	43.9±6.61 ^{f}
Sperm concentration (x 10^6/ml)	69.1±6.46 ^{a}	16.7±1.51 ^{b}	56.7±10.33 ^{c}	51±9.15 ^{d}	15.3±2.55 ^{b}	50.6±10.63 ^{d}
Progressive motility percent(A+B)%	65.1±5.99 ^{a}	57.5±5.3 ^{b}	38.73±3.9 ^{c}	57±6.68 ^{b}	26.7±7.81 ^{d}	32.3±6.18 ^{e}
Normal morphology percent %	58.5±7.76 ^{a}	44.8±8.2 ^{b}	42.8±5.9 ^{c}	24±3.75 ^{d}	42.45±6.62 ^{c}	22.4±3.29 ^{e}
Leukocyte concentration (x 10^6/ml)	1.66±0.76 ^{a}	2.5±0.85 ^{b}	2.7±0.83 ^{b}	3.3±0.65 ^{c}	3.8±0.71 ^{d}	4.37±1.29 ^{e}
MDA concentration (µmol/l)	3.7±0.44 ^{a}	5.52±1.5 ^{b}	6.52±0.89 ^{c}	7.6±1.09 ^{d}	8.1±1.11 ^{e}	8.6±1.65 ^{f}
Sperm chromatin abnormality percent%	32.1±3.09 ^{a}	35±3.8 ^{b}	38±3.95 ^{c}	40±4.44 ^{d}	42.6±4.18 ^{e}	45±6.18 ^{f}

Value are expressed as mean±S.D.

different letters means significant difference ($P<0.05$).

UXI: Unexplained infertility , O: Oligozoospermia , A: Asthenozoopermia , T: Teratozoopermia,

OA: Oligoasthenozoopermia ,AT: Asthenoteratozoospermia

4-2: Effect of invitro activation on spermiogram of unexplained semen using swim – up and density gradient centrifugation techniques.

The results of certain sperm parameters namely: sperm concentration, progressive motility percent and normal morphology percent as well as leukocyte concentration, MDA concentration and sperm chromatin abnormality percent following invitro sperm activation by swim – up and density gradient techniques of unexplained semen specimens were shown in table (4-2).

No significant ($P > 0.05$) difference was recorded between two semen specimens before activation in all sperm parameters as well as leukocyte concentration, MDA concentration and sperm chromatin abnormality percent.

There was a significant ($P < 0.05$) decrease in the mean of sperm concentration after activation by both methods compared to before activation while no significant ($P > 0.05$) difference was observed between swim – up and density gradient techniques following activation.

The results in this table showed a significant ($P < 0.05$) elevation in progressive motility (A + B) and normal morphology percent after activation by swim – up and density gradient techniques when compared with before activation. A significant ($P < 0.05$) increase in progressive motility (A + B) by using density gradient technique compared to swim – up technique whereas no significant ($P > 0.05$) difference appeared between two methods following activation in normal morphology percent.

On the other hand, there was a significant ($P < 0.05$) reduction in the leukocyte concentration and MDA concentration as well as sperm chromatin abnormality percent after activation by both methods as compared with before activation while a significant ($P < 0.05$) decrease was showed by using density gradient technique compared to swim – up technique.

Table(4-2) Effect of invitro activation on spermiograms of unexplained semen using swim-up and density gradient centrifugation techniques

Patients / Variables	Swim – up		Density gradient	
	Before Activation n=15	After Activation n=15	Before Activation n=15	After Activation n=15
Sperm concentration (x 10^6/ml)	68 ± 5.94^{a}	33.3 ± 3.47^{b}	70.1 ± 6.98^{a}	28.7 ± 4.47^{b}
Progressive motility percent(A+B)%	64.8 ± 6.24^{a}	73.6 ± 6.78^{b}	65.3 ± 5.95^{a}	84.3 ± 7.44^{c}
Normal morphology percent %	59 ± 5.94^{a}	67.3 ± 5.15^{b}	57.4 ± 9.42^{a}	70.7 ± 6.35^{b}
Leukocyte concentration (x 10^6/ml)	1.62 ± 0.67^{a}	1.04 ± 0.3^{b}	1.7 ± 0.87^{a}	0.61 ± 0.41^{c}
MDA concentration (µmol/l)	3.7 ± 0.35^{a}	2.6 ± 0.49^{b}	3.6 ± 0.51^{a}	1.8 ± 0.23^{c}
Sperm chromatin abnormality percent%	31.5 ± 2.97^{a}	22.3 ± 1.58^{b}	32.7 ± 3.19^{a}	17.3 ± 1.13^{c}

Value are expressed as mean ±S.D.

different letters means significant difference ($P<0.05$).

4-3: Effect of in vitro activation on spermiogram of oligozoospermic semen using swim–up and density gradient centrifugation techniques.

Table (4-3) revealed no significant ($P > 0.05$) difference between two semen specimens before activation in all sperm parameters as well as leukocyte concentration, MDA concentration and sperm chromatin abnormality percent.

There was a significant ($P < 0.05$) decline in mean sperm concentration after activation by both methods compared to the results before activation and no significant ($P > 0.05$) difference was noticed when compared between two techniques.

As well as a significant ($P < 0.05$) improvement appeared in progressive motility (A + B) and normal morphology percent following activation by both methods as compared with before activation. Moreover, there was a significant ($P < 0.05$) elevation in progressive motility (A + B) by using density gradient technique compared to swim – up technique while no significant ($P > 0.05$) difference was noticed in normal morphology percent between two methods after activation. Lastly, the results in table (4-3) recorded a significant ($P < 0.05$) reduce in the leukocyte concentration and MDA concentration as well as sperm chromatin abnormality percent following activation by swim – up and density gradient techniques as compared with before activation and also a significant ($P < 0.05$) decrease appeared by using density gradient technique compared to swim – up technique.

Table(4-3) Effect of invitro activation on spermiograms of oligozoospermic semen using swim-up and density gradient centrifugation techniques.

Patients / Variables	Swim – up		Density gradient	
	Before Activation n=15	After Activation n=15	Before Activation n=15	After Activation n=15
Sperm concentration (x 10^6/ml)	16.46±1.55 ^{a}	7.1±0.83 ^{b}	16.9±1.48 ^{a}	7.6±0.89 ^{b}
Progressive motility percent(A+B)%	57.3±5.65 ^{a}	69.9±5.06 ^{b}	57.8±5.28 ^{a}	79.1±7.66 ^{c}
Normal morphology percent %	42.9±8.76 ^{a}	64.6±8.73 ^{b}	46.7±7.55 ^{a}	65.9±5.35 ^{b}
Leukocyte concentration (x 10^6/ml)	2.5±1.04 ^{a}	1.8±0.93 ^{b}	2.5±0.63 ^{a}	1.2±0.52 ^{c}
MDA concentration (µmol/l)	5.9±1.74 ^{a}	4.3±1.38 ^{b}	5.1±1.18 ^{a}	3.2±0.97 ^{c}
Sperm chromatin abnormality percent%	35.7±2.37 ^{a}	27.5±2.47 ^{b}	34.5±3.09 ^{a}	22.2±2.55 ^{c}

Value are expressed as mean ±S.D.

different letters means significant difference ($P<0.05$).

4-4: Effect in vitro activation on spermiogram of asthenozoopermic semen using swim – up and density gradient centrifugation techniques.

According to the results in table (4-4) there was no significant ($P > 0.05$) difference between two semen specimens before activation in all sperm parameters as well as leukocyte concentration, MDA concentration and sperm chromatin abnormality percent.

Activation asthenozoopermic semen samples using swim – up and density gradient methods revealed a significant ($P < 0.05$) decrease in the mean of sperm concentration when compared with before activation while no significant ($P > 0.05$) difference appeared between two techniques following activation.

The results were appeared a significant ($P < 0.05$) improved in progressive motility (A + B) as well as normal morphology percent after activation by swim – up and density gradient techniques when compared with before activation and then a significant ($P < 0.05$) increase was showed in progressive motility (A + B) by using density gradient technique compared to swim – up technique while no significant ($P > 0.05$) difference appeared in normal morphology percent between two methods following activation.

Concerning the resulting values of leukocyte concentration, MDA concentration and sperm chromatin abnormality percent caused a significant ($P < 0.05$) reduction after activation by both methods as compared with before activation. Moreover, there was a significant ($P < 0.05$) decrease in leukocyte concentration by using density gradient technique compared to swim – up technique whereas no significant ($P > 0.05$) difference was noticed between two methods following activation in MDA concentration and sperm chromatin abnormality percent.

Table(4-4) Effect of invitro activation on spermiograms of asthenozoopermic semen using swim-up and density gradient centrifugation techniques.

Patients / Variables	Swim – up		Density gradient	
	Before Activation n=15	After Activation n=15	Before Activation n=15	After Activation n=15
Sperm concentration (x 10^6/ml)	56.2±10.83 [a]	26.6±6.27 [b]	57.3±10.1 [a]	27.1±6.77 [b]
Progressive motility percent(A+B)%	37.6±3.53 [a]	65±7.45 [b]	39.9±4.08 [a]	70.6±6.93 [c]
Normal morphology percent %	41.8±5.21 [a]	56.5±6.50 [b]	43.7±6.57 [a]	58.7±7.84 [b]
Leukocyte concentration (x 10^6/ml)	2.8±0.74 [a]	2.03±0.65 [b]	2.6±0.92 [a]	1.4±0.53 [c]
MDA concentration (µmol/l)	6.5±0.84 [a]	4.5±0.62 [b]	6.6±0.94 [a]	3.9±0.62 [b]
Sperm chromatin abnormality percent%	38.7±4.19 [a]	28.1±3.84	37.4±3.71 [a]	26.5±4.07 [b]

Value are expressed as mean ±S.D.

different letters means significant difference ($P<0.05$).

4-5: Effect in vitro activation on spermiogram of teratozoospermic semen using swim – up and density gradient centrifugation techniques.

Concerning the results in the table (4-5), it was found no significant (P > 0.05) difference between two semen specimens before activation in all sperm parameters as well as leukocyte concentration, MDA concentration and sperm chromatin percent.

Table (4-5) revealed a significant (P < 0.05) decrease in the mean of sperm concentration following activation by swim – up and density gradient techniques as compared with before activation. Whereas, no significant (P > 0.05) difference between two methods after activation.

There was a significant (P < 0.05) elevation in progressive motility (A + B) as well as the percentage of normal morphology following in vitro activation by both methods compared to before activation. While no significant (P > 0.05) difference was recorded between two techniques in these portions. Recording the results of leukocyte concentration and MDA concentration as well as the percentage of sperm chromatin abnormality was showed in this table a significant (P < 0.05) reduction after activation by both methods when compared with results before activation and no significant (P > 0.05) was recorded in this variables between two methods following activation.

Table(4-5) Effect of invitro activation on spermiograms of teratozoospermic semen using swim-up and density gradient centrifugation techniques.

Patients / Variables	Swim – up		Density gradient	
	Before Activation n=12	After Activation n=12	Before Activation n=13	After Activation n=13
Sperm concentration (x 10^6/ml)	50.5±7.76 [a]	26.2±5.93 [b]	51.7±6.93 [a]	26.8±5.10 [b]
Progressive motility percent(A+B)%	56±4.43 [a]	64.4±4.01 [b]	58±6.67 [a]	66.1±6.96 [b]
Normal morphology percent %	25.1±3.75 [a]	47.3±6.12 [b]	23±3.59 [a]	51.3±7.06 [b]
Leukocyte concentration (x 10^6/ml)	3.2±0.56 [a]	2.2±0.72 [b]	3.4±0.68 [a]	1.9±0.52 [b]
MDA concentration (µmol/l)	7.2±1.20 [a]	5.8±1.27 [b]	8.1±0.76 [a]	5.4±0.56 [b]
Sperm chromatin abnormality percent%	40.5±5.53 [a]	31.2±5.19 [b]	39.6±3.02 [a]	28.6±3.91 [b]

Value are expressed as mean ±S.D.

different letters means significant difference ($P<0.05$).

4-6: Effect in vitro activation on spermiogram of oligoasthenozoopermic semen using swim – up and density gradient centrifugation techniques.

Table (4-6) found no significant ($P > 0.05$) difference between two semen specimens before activation in all sperm parameters as well as leukocyte concentration, MDA concentration and sperm chromatin abnormality percent.

There was a significant ($P < 0.05$) decline in the sperm concentration, leukocyte concentration, MDA concentration and sperm chromatin abnormality percent following in vitro activation by both methods compared to before activation. Whereas, no significant ($P > 0.05$) difference appeared between two techniques.

Lastly, the results of progressive motility (A + B) and normal morphology percent showed in this table a significant ($P < 0.05$) increase after activation by swim – up and density gradient methods when compared with before activation while no significant ($P > 0.05$) difference was showed between two methods following activation.

Table(4-6) Effect of invitro activation on spermiogram: of oligoasthenozoopermic semen using swim-up and density gradient centrifugation techniques.

Patients / Variables	Swim – up		Density gradient	
	Before Activation n=10	After Activation n=10	Before Activation n=10	After Activation n=10
Sperm concentration (x 10^6/ml)	14.9±2.84 [a]	7.7±1.56 [b]	15.7±2.31 [a]	7.4±1.57 [b]
Progressive motility percent(A+B)%	25.8±4.99 [a]	54.2±6.46 [b]	27.7±5.13 [a]	56±7.11 [b]
Normal morphology percent %	41.9±6.87 [a]	55.3±7.43 [b]	43±6.69 [a]	56.2±7.3 [b]
Leukocyte concentration (x 10^6/ml)	3.7±0.39 [a]	2.6±0.79 [b]	3.8±0.95 [a]	2.3±0.37 [b]
MDA concentration (µmol/l)	7.9±1.06 [a]	5.8±1.06 [b]	8.4±1.17 [a]	5.6±0.67 [b]
Sperm chromatin abnormality percent%	43.4±3.77 [a]	34.6±3.85 [b]	41.9±4.62 [a]	32.6±4.37 [b]

Value are expressed as mean±S.D.

different letters means significant difference ($P<0.05$).

4-7: Effect of in vitro activation on spermiogram of asthenoteratozoospermic semen using swim – up and density gradient centrifugation techniques.

In table (4-7), it was found no significant ($P > 0.05$) difference between two semen specimens before activation in all sperm parameters as well as leukocyte concentration, MDA concentration and sperm chromatin abnormality percent.

A significant ($P < 0.05$) reduction was revealed in sperm concentration and leukocyte concentration as well as MDA concentration and sperm chromatin abnormality percent following activation by both methods when compared with before activation but no significant ($P > 0.05$) difference was observed between swim – up technique compared to density gradient techniques.

On other hand, the results of progressive motility (A + B) and normal morphology percent showed a significant ($P < 0.05$) elevation following in vitro activation by both methods compared to before activation while no significant ($P > 0.05$) difference was noticed between two techniques after activation.

Table(4-7) Effect of invitro activation on spermiograms ، f asthenoteratozoospermic semen using swim-up and density gradient centrifugation techniques.

Patients / Variables	Swim – up		Density gradient	
	Before Activation n=10	After Activation n=10	Before Activation n=10	After Activation n=10
Sperm concentration (x 10^6/ml)	51.5 ±10.78 [a]	28.6±6.93 [b]	49.8±11 [a]	27.7±7.11 [b]
Progressive motility percent(A+B)%	31.3±6.37 [a]	52.4±8.23 [b]	33.3±6.14 [a]	54±7.03 [b]
Normal morphology percent %	21.3±3.33 [a]	46.1±6.40 [b]	23.5±3.02 [a]	49.8±7.53 [b]
Leukocyte concentration (x 10^6/ml)	4.2±1.18 [a]	3.2±0.80 [b]	4.5±1.45 [a]	2.9±1.04 [b]
MDA concentration (µmol/l)	9.01±1.21 [a]	6.2±0.88 [b]	8.3±1.60 [a]	5.5±1.21 [b]
Sperm chromatin abnormality percent%	45.3±4.69 [a]	38.6±3.53 [b]	44±7.64 [a]	35.6±6.14 [b]

Value are expressed as mean ±S.D.

different letters means significant difference ($P<0.05$).

4.8: Prognostic factors affecting on the IUI outcome of infertile patients.

Figure (4-1) IUI outcome showed a significant (p <0.01) difference among subgroups of infertile patients. The high value of pregnancy rate was showed with unexplained group (20%), then Oligozoospermia (10%), Asthenozoopermia (6.7%) and teratozoospermia (4%). While less value was noticed with oligoasthenozoopermia and asthenoteratozoospermia (0%).

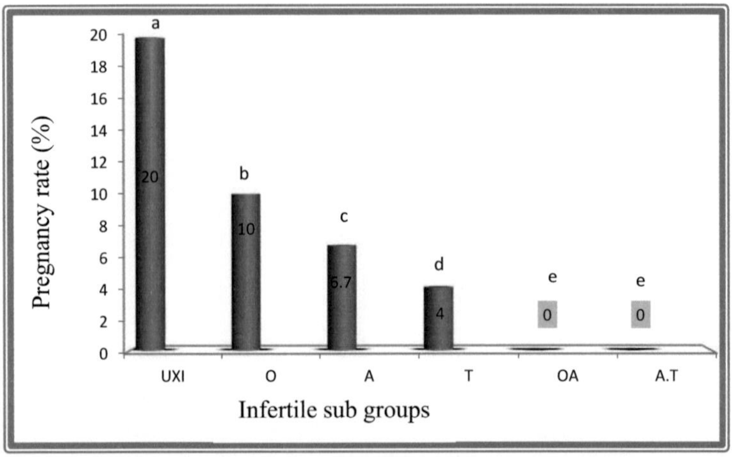

Figure (4-1) IUI outcome according sub groups of infertile patients after activation by swim – up and density gradient techniques

UXI : unexplained infertility (n = 30)

O : Oligozoospermia (n = 30)

A : Asthenozoospermia (n = 30)

T : Teratozoospermia (n = 25)

O.A : Oligoasthenozoopermia (n = 20)

A.T: Asthenoteratozoospermia (n = 20)

Different letters means significant difference (P < 0.01)

The results in figure (4-2) showed a significant (P < 0.01) increase in the pregnancy rate after activation by density gradient method (9.09%) when compared with swim – up method (6.4%).

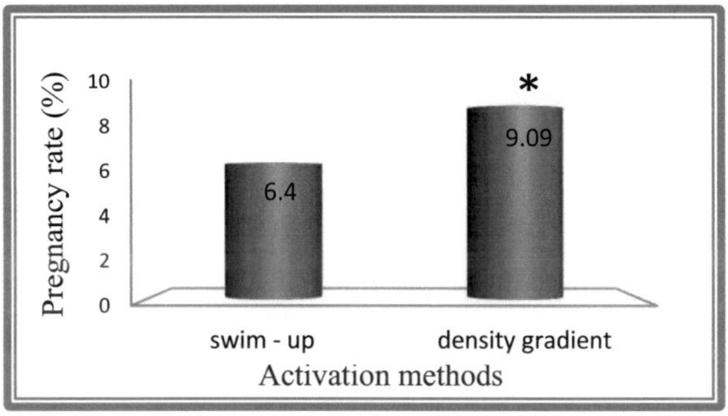

Figure (4-2) comparison of IUI outcome for infertile patients after activation between swim – up and density gradient techniques

∗ means significant difference (P < 0.01)

Figure (4-3) there was highly significant (P < 0.01) improvement was observed in pregnancy rate of unexplained patients after activation by using density technique (26.7%) compared with swim – up technique (13.3%).

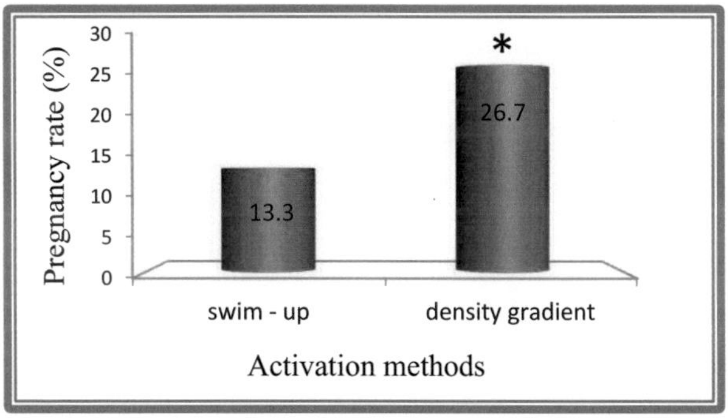

Figure (4-3) comparison of IUI outcome for unexplained patients after activation between swim – up and density gradient techniques.

∗ means significant difference (P < 0.01).

Figure (4-4) showed a highly significant ($P < 0.01$) elevation in pregnancy rate of oligozoospermic patients after activation by using density gradient technique (13.3%) compared with swim – up technique (6.7%).

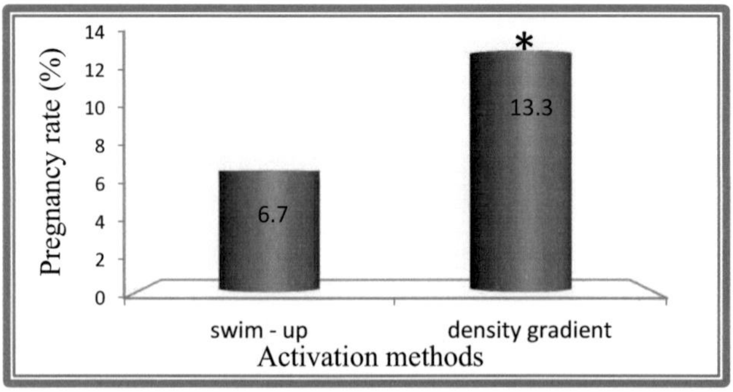

Figure (4-4) comparison of IUI outcome for oligozoospermic patients after activation between swim – up and density gradient techniques

∗ means significant difference ($P < 0.01$)

Figure (4-5) There was no significant ($P > 0.01$) difference in pregnancy rate of asthenozoopermic patients after activation between swim – up (6.7%) and density gradient (6.7%) techniques.

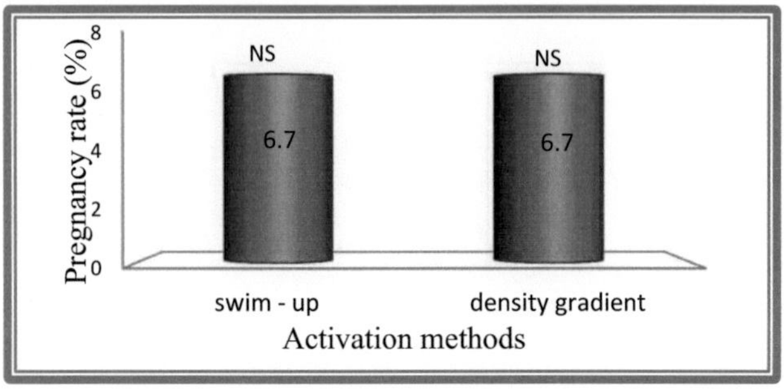

Figure (4-5) comparison of IUI outcome for asthenozoopermic patients after activation between swim – up and density gradient techniques

NS: no significant difference ($P > 0.01$)

The results in figure (4-6) recorded a highly significant (P < 0.01) increase in pregnancy rate of teratozoospermic patients after activation by using density gradient technique (8.3%) compared with swim – up technique (0%).

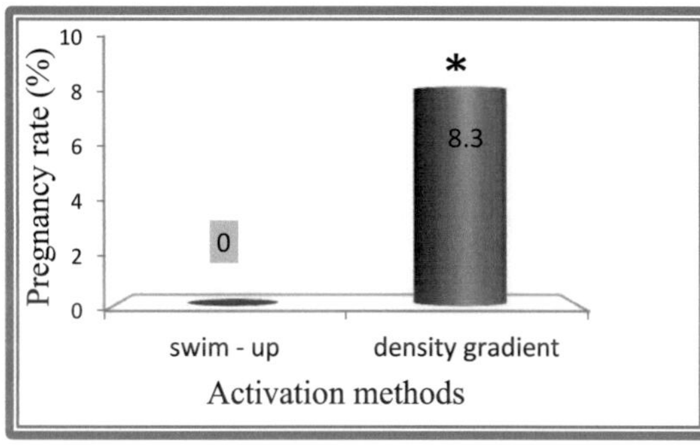

Figure (4-6) comparison of IUI outcome for teratozoospermic patients after activation between swim – up and density gradient techniques

∗ means significant difference (P < 0.01)

Figure (4-7) No pregnancy was observed in Oligoasthenozoopermic group in both methods.

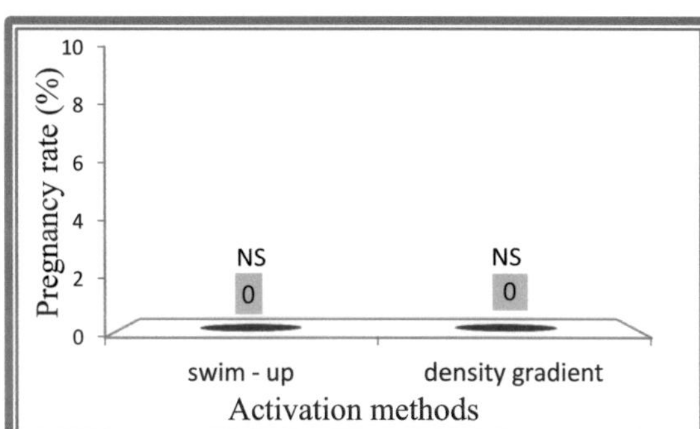

Figure (4-7) comparison of IUI outcome for Oligoasthenozoopermia patients after activation between swim – up and density gradient techniques

NS: no significant difference (P > 0.01).

The results in figure (4-8) were appeared no pregnancies was observed in asthenoteratozoospermic group in both techniques .

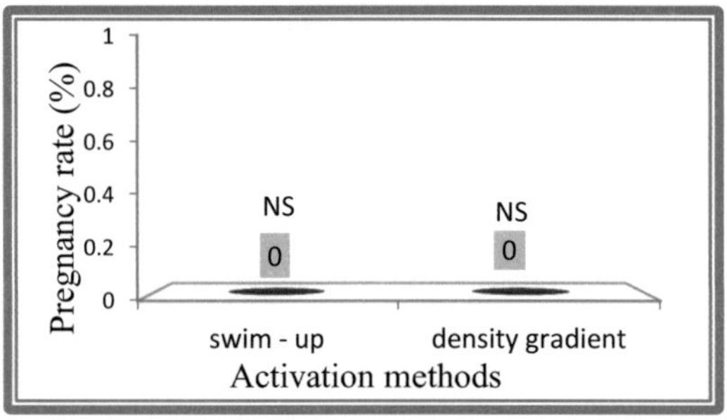

Figure (4-8) comparison of IUI outcome for asthenoteratozoospermic patients after activation between swim – up and density gradient techniques

NS: no significant difference (P > 0.01).

Pregnancy rate according to sperm concentration is shown in figure (4-9) the highest pregnancy rate was achieved among couples with sperm concentration more than 40 million (10.9%) followed by those with sperm concentration between 20-40 million (7.5%), while the pregnancy rate was observed in those with sperm concentration of less than 20 million (5%) .

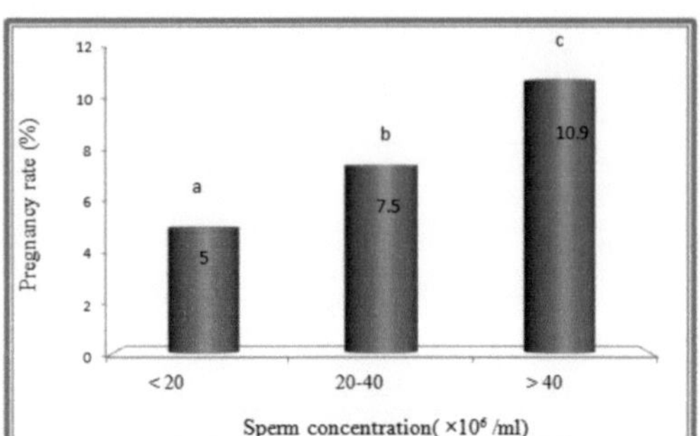

Figure (4-9) Effect of the sperm concentration on the IUI outcome for infertile patients .

Different letters means significant difference (P < 0.01)

Our results in the figure (4-10) showed that the pregnancy rate increased with couples that had 50% or greeter of progressive motility percent (A + B) (10%) compared to less than 50% (5.8%).

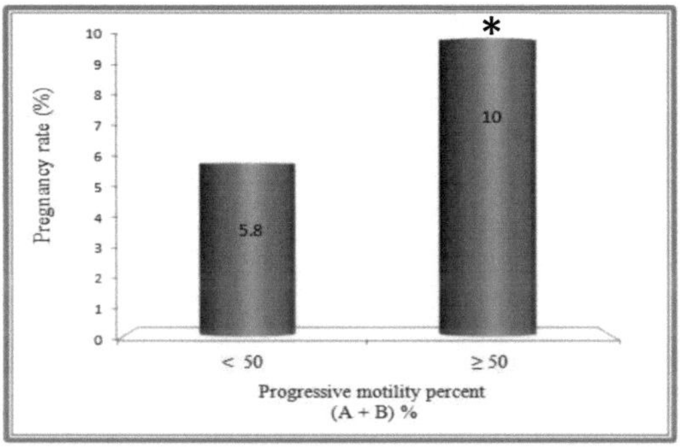

Figure (4-10) Effect of the progressive motility percent on the IUI outcome of infertile patients .

* means significant difference (P < 0.01)

In the figure (4-11 there was statistically significant (P < 0.01) increase in pregnancy rate among couples in whom normal morphology percent was ≥ 30% (10%) as compared to those with < 30% (5.3%).

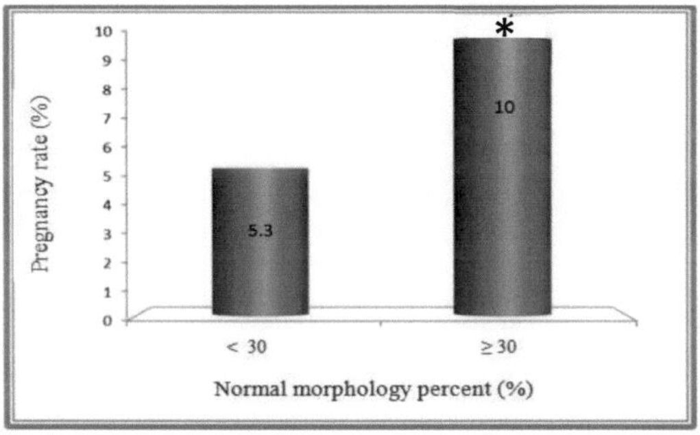

Figure (4-11) Effect of the normal morphology percent on the IUI outcome of infertile patients .

* means significant difference (P < 0.01)

Figure (4-12) The pregnancy rate was better in couples with leukocyte concentration less than 1 million (16.6%) followed by those with leukocyte concentration between 1-2 million (10%) and more than 2-5 million (5.5%), while the pregnancy rate showed higher decline in those with leukocyte concentration more than 5 million (3.6%).

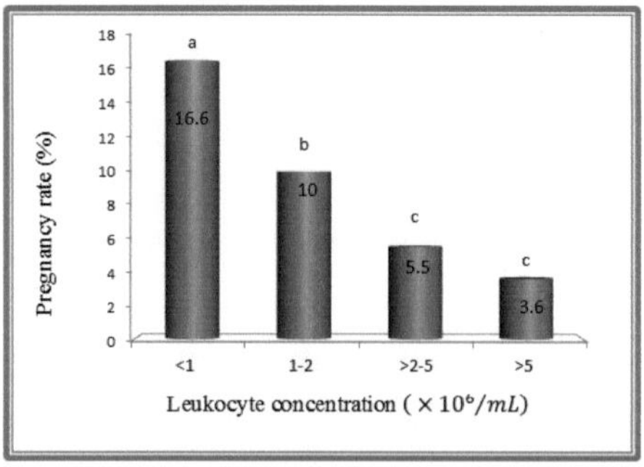

Figure (4-12) Effect of the leukocyte concentration on the IUI outcome of infertile patients .

Different letters means significant difference ($P < 0.01$)

The pregnancy rate according to the MDA concentration was observed in figure (4-13) The pregnancy rate with couples who have less than 4 μmol/L was significantly higher than ≥ 4 μmol/L or greeter (11.4% versus 4.7%).

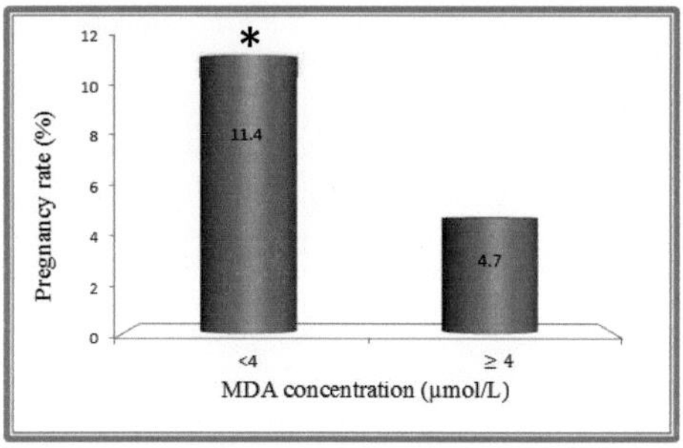

Figure (4-13) Effect of the MDA concentration on the IUI outcome of infertile patients .

∗ means significant difference ($P < 0.01$)

In the figure (4-14) The pregnancy rate was much higher in the couples that have less than 30% of sperm chromatin abnormality percent (10.7%), while the couples that have 30% or greater had the lowest pregnancy rate (5.5%).

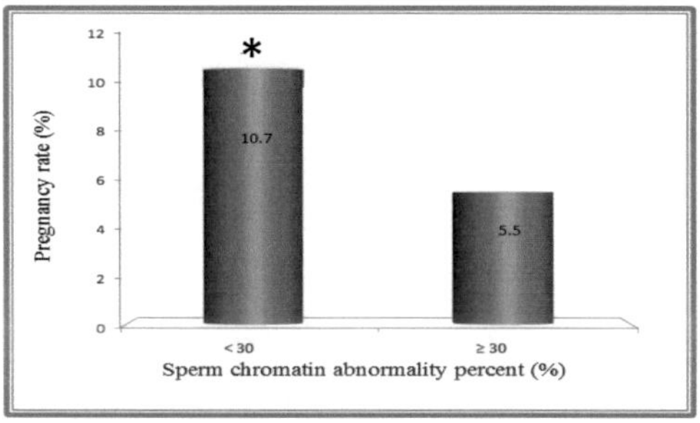

Figure (4-14) Effect of sperm chromatin abnormality percent on the IUI outcome of infertile patients .

∗ means significant difference (P < 0.01)

Figure (4-15) The pregnancy rate was reduced when the couples age increased. For less than 30 years, the pregnancy rate higher significant (16.6%) followed by those with age between 30-34 years (8%) and age between 35-39 years (5.2%), while the pregnancy rate significantly reduced with age more than 40 years (4.8%) .

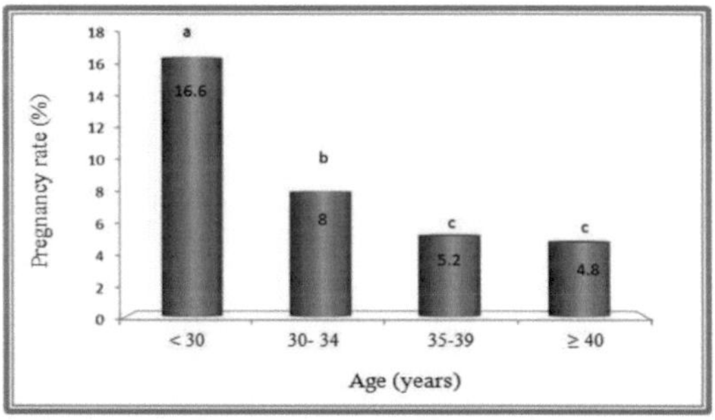

Figure (4-15) Effect of the age on the IUI outcome of infertile patients .

Different letters means significant difference (P < 0.01)

Figure (4-16) showed no statistical significant was observed in the pregnancy rate for type of infertility between secondary infertility (7.8%) and primary infertility (7.5%).

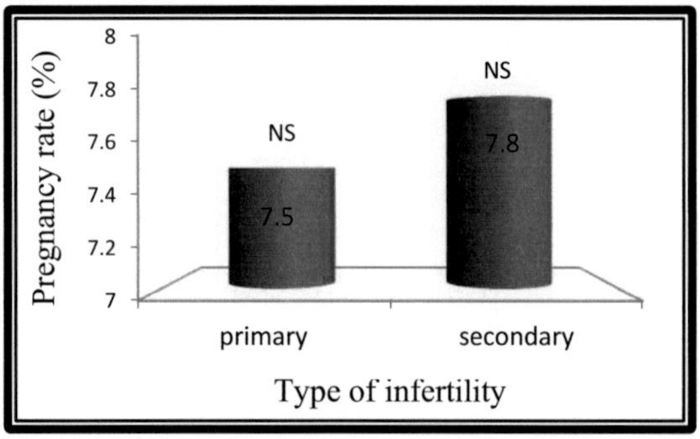

Figure (4-16) Effect type of infertility on the IUI outcome of infertile patients.
NS: no significant difference (P > 0.01)

Difference was statistically significant at (P < 0.01). in the figure (4-17) we found that 4-6 years duration of infertility had the highest pregnancy rate (10%). The pregnancy rate for duration between 1-3 years was (8%). The pregnancy rate reduced to (4%) after 6 years of infertility.

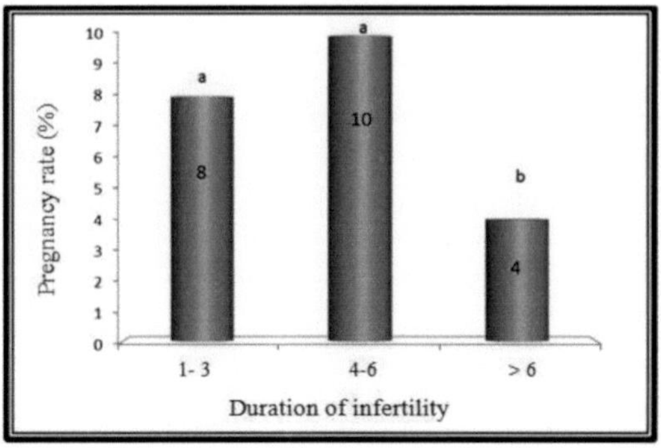

Figure (4-17) Effect duration of infertility on the IUI outcome of infertile patients .

Different letters means significant difference (P < 0.01)

Figure (4-18) the pregnancy rate of the non – smoker was showed a significantly (P < 0.01) higher than those of the smoker (9.2% versus 5.8%) .

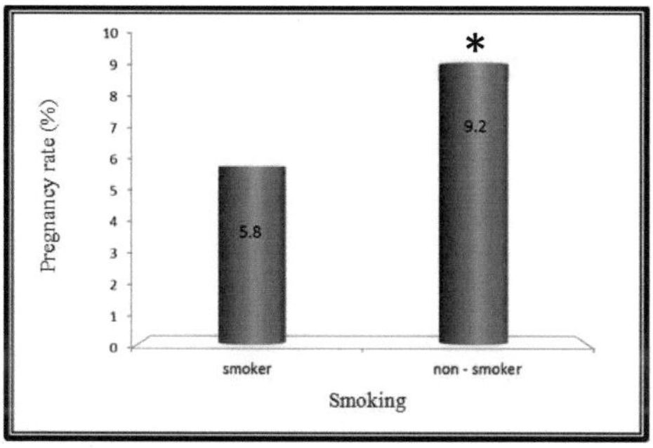

Figure (4-18) Effect smoking on the IUI outcome of infertile patients .

∗ means significant difference (P < 0.01).

Discussion

5-1: Effect of semen parameters, seminal MDA level and sperm chromatin abnormality percent for infertile patients:

Defective sperm parameters indicate an important role in diagnosis fertility and was closely correlated to fertilization and pregnancy rate in the natural fertilization process as well as in ART (Burger, 2013).

As shown in (table 4-1) there is a significant improvement ($P < 0.05$) in semen volume, sperm concentration, progressive motility and normal morphology for UXI compared to five abnormal semen groups including (O, A, T, OA and AT respectively) while the results in the same table showed a significantly decrease ($P < 0.05$) in liquefaction time, leukocytes concentration, MDA concentration and sperm chromatin abnormality for UXI compared with abnormal semen groups. This may be due to the increased concentrations of leukocytes in semen which reports the presence of a genital tract infection . It is associated with an increase in immature germ cell concentration.

The effects of leukocytospermia on sperm may lead to increased oxidative stress and DNA damaged that are demonstrated in various studies. The production of ROS causing defect of sperm chromatin/high DNA damage can be attributed to poor sperm quality and lower pregnancy rate following ART (Menkveld, 2004).

Saleh and Agarwal (2002) have suggested that leukocytes is the main source of oxidative stress in seminal plasma, this is because ROS at high values are toxic to sperm quality and function. Various studies were indicated that oxidative stress induce damage of sperm DNA via defective of sperm chromatin and break strands of DNA (Aitken and Baker, 2006).

leukocytospermia are also associated with increased levels of sperm DNA damage, especially in abnormal ejaculation patients. Oxidative stress as main cause of sperm DNA defect with others aspects like apoptosis, sperm chromatin remodeling and defective packaging processes (Venkatesh *et al.* , 2011).

In the present study, significant higher values of MDA levels and sperm chromatin abnormality together with poor semen quality in the raw semen samples of leukocytospemic patients may indicate that lipid peroxidation has made adverse alter in sperm membrane that leads to failed sperm motility, morphology and sperm concentration, its clear that lipid peroxidation may change sperm membrane function and cause abnormal sperm quality.

The results of the present study showed an agreement with many previous studies that revealed the positive correlation between DNA damage and oxidative stress which is associated with failed sperm production, this may be due to the production of immature and abnormal form of spermatozoa and has a defective sperm chromatin. (Sakkas *et al.*, 1999). Their study showed a significant increase in MDA concentration of infertile patients, therefore, lipid peroxidation attack and degredate the membranes of sperm and causing abnormal sperm quality and function (Ben abdallah *et al.*, 2009). Some studies appeared the lipid peroxidation negatively associated with potential sperm parameters between control and infertile patients (Agarwal and Allamaneni, 2011 ; Hsieh *et al.*, 2006).

Regarding the sperm concentration, many studies have suggested clear relatedness between abnormal chromatin and sperm count which reflect the impact of protamine deficiency in cousing apoptosis or impaired spermatogenesis (Lakpour *et al.*, 2012; Sakkas *et al.*, 1999). Moreover , sperm chromatin defect involve the increase production of OS which may enhance the processes of apoptosis that tend to reduce sperm concentration that is related with sub infertility (Agarwal and allamaneni, 2004). Some investigators suggest that early apoptosis is related with immature nuclear development and then abnormal nuclear decondensed chromatin (Said *et al.*, 2006).

In relation to sperm motility, the results in the present study showed in agreement with many studies have found that excessive level of OS is depleted adenosine triphosphate (ATP) which rapidly lead to decrease proteins phosphorylation causing transient failure of motility (Armstrong *et al.*, 1999; De Lamirande and Cagon, 1992). Also some researchers indicated sperm chromatin decondensation which association degrees with motility (Lakpour *et al.*, 2012; Kazerooni *et al.*, 2009). Hamad (2014) who found that infertile men with protamine deficiency display Lower sperm counts, motility and higher abnormal sperm morphology. In the opposite, Sellami *et al.* (2013) demonstrated that there is no correlation between sperm chromatin condensation assessed by aniline blue staining and some parameters, including motility, vitality and sperm count. This comes in agreement with other studies that did not find any correlation between sperm chromatin and sperm parameters (Sepaniak *et al.*, 2006; Henkel *et al.*, 2004).

Concerning to the semen volume, abnormal synthesis and secretion in accessory sex glands may reflect low semen volume (Siegel, 1993). The increase of leukocyte concentration , especially monocyte in the infertile patient which may cause low semen volume (Wolff *et al.*, 1990). Aleksandra *et al.* (2004) which proposed that an increase of leukocyte concentration may lead to oxidative stress which is related with decrease semen volume making decline sperm quality and male infertility.

Regarding the liquefaction time, male accessory glands contain potential proteins which an important to the coagulation and liquefaction of semen.

Abnormal viscosity of semen reflects hypo function of the seminal vesicles or prostate, therefore hyper viscous seminal fluid which may result because of infection and high levels of seminal leukocytes (Du Plessis *et al.*, 2013).

Oxidative damage may play a role in semen hyper viscous during excessive ROS production by increase amount of seminal leukocyte. (Alkan *et al.*, 1997).

5-2: Effect in vitro activation on spermiograms of infertile patients using swim – up and density gradient techniques

Sperm activation techniques play major role in the improvement of sperm ability of fertilization for infertile patients before artificial insemination (Donnelly *et al.*, 2000). Therefore, Henkel and Schill (2003) reported these activation techniques for intrauterine insemination which increase the number of motile sperm and normal morphology, and to eliminate damage of immotile sperm, leukocytes, bacteria, debris and other cells that yield ROS that negatively have effect on the fertilization and pregnancy rate.

The sperm concentration was significantly reduced after activation by both methods. This due to that non – active or immotile sperm which remains in the bottom of activation tube and only sperm have excellent motility and normal morphology which are able to swim – up to upper layer. While, the density gradient technique has separated cells of seminal fluid according to density, therefore sperm move actively and normal form can reach to pellet while the immotile sperm and other cells will reach to the upper layer of activation tube. The same observation was showed by several studies (Al-Dujaily and Malik, 2013; WHO, 2010).

Improvement of sperm motility occurs because the culture media contain special components which lead to the increased number of motile sperm and normal form during activation (Schlegel and Girardi, 1997). The benefit of invitro culture media is to decrease the viscosity of the seminal fluid and promote sperm move more freely and finally increase sperm motility and grade activity (Makler *et al.*, 1984).

Also, culture media reduce the damage obtained by seminal fluid substituents as leukocyte, bacteria, antisperm antibodes and others which causes decrease progressive motility (Makler and Jacobi, 1981). Furthermore,

Kaewnoonual *et al.*, (2008) reported the different media which play an important role to assist the improvement of semen quality and successful recovery of infertile patients. Therefore, sperm activation techniques should use culture media that involve buffers and protein source to promote sperm hyperactivation and capacitation, also adding glucose, pyruvate and sodium lactate to generate energy (Baker *et al.*, 2000).

The percentage of normal morphology showed significantly improvement by both methods after activation. This may be eliminate leukocytes, round cells, debris and others that lead to reduce ROS production that induce DNA damage . Therefore, normal spermatozoa moves fast from seminal plasma into layer of culture media (Sharma *et al.*, 2004). In the same portion, Mortimer (2000) proposed the decapacitation factors in the seminal plasma that inhibit spontaneous capacitation, which has an impact on sperm function. Therefore recovery by culture media for infertile semen during preparation techniques has been improved morphological spermatozoa and progressive motility that are required for ART. This agrees with Henkel and Schill (2003) who reported that sperm activation techniques used for infertile semen samples to increase motility and normal morphology through decrease levels of ROS.

Increase seminal leukocytes particularly macrophage and neutrophils that stimulate male subfertility through damaging sperm by the yield of ROS and apoptosis (Tremellen and Tunc, 2010) Aziz *et al.*, (2004) reported a significant reduce in sperm motility which was correlated with increase of leukocyte concentration. Normal sperm function requires a balance between the ROS produced during sperm maturation process in the epididymis and the antioxidants levels found in the secretions of the genital male tract. (El-Tohamy,2012).

Consequently culture media contain essential components which may maintain sperm membranes from effect of ROS by generating adenosine triphosphate (ATP) during transport fatty acids into the mitochondria therefore DNA damage is reduced (Berni *et al.*, 2008).

On the other hand, sperm activation and assisted reproductive techniques assist to reduce the effect of ROS, therefore , the presence antioxidants in the culture media decrease ROS production is by the immature sperm and leukocyte and increase fusion of sperm-oocyte (Irvine, 1996). Also Aitken *et al.*, (1992) reported the improvement in the percentage of normal morphology after in vitro activation of different groups of male semen samples. This due to the decrease of the major source of excess ROS production in seminal plasma.

Data of the current study noted an improvement of sperm chromatin abnormality following activation by both swim – up and density gradient techniques with increase of normal morphology and progressive motility. This may be due to the effect of both techniques (swim – up and density gradient) which are used to the isolation of normal sperm from abnormal as well as removal leukocyte which is the major source of ROS . All of these will given good results by reduction the percentage of chromatin abnormality. Also other causes that could impact positively the improvement in the percentage of sperm chromatin abnormality in this study. This is may be various dilutions obtained through activation leading to absent in leukocyte and decreased abnormal cells and other cells which may yield ROS.

This finding agrees with many studies that proposed activation by density gradient and swim – up that increase sperm quality with normal percentage of sperm chromatin (Hammadeh *et al.*, 2001; Sanchez *et al.*, 1994). Adham *et al.* (2001) concluded the relationships between chromatin condensation and morphological alteration of the acrosomal vesicle through spermatogenesis because impairment of condensation leads to failure association of acrosome to the nucleus. Also these results are similar with our studies, which considered leukocyte and immature spermatozoa potential source for ROS production (Henkel *et al.*, 2005; Baker *et al.* , 2003). Compact organization and antioxidants in semen normally maintain the sperm DNA. DNA repair based on the oocyte for repairing following fertilization (Aitken *et al.*, 2003). Liu (2013) indicated presence of greater negative relation between normal morphology percent and quality of sperm chromatin packaging. Therefore assessment of clinical validity of sperm DNA to increase the positive impact of two preparation techniques and motility that promote and add information for sperm quality that is required in ART outcome (Schulte *et al.*, 2010).

The current study used two techniques (swim – up and density gradient) which were given more effectiveness in certain sperm parameters and were improved through activation by density gradient technique more than by swim – up technique. These data are similar to a study by Boomsma *et al.* (2007) which suggested the density gradient technique presented more benefit in sperm treatment than swim – up technique, density gradient method provide greater number of motile sperm and decrease the contamination of bacteria and procedure time. Moreover, density gradient method has given a greater percentage of normal morphology of sperm treatment with best DNA integrity and chromatin maturity (Morrell *et al.*, 2004).

Therefore, density gradient and various activation methods have significantly decrease ROS production by elimination leukocytes (Makker *et al.*, 2009).

Many studies that indicated the recovery techniques of sperm such as density gradient and swim – up used to isolate of functional sperm from non – functional during activation leading to elevation in the percentage of normal morphology of sperm for all groups of male semen samples with higher percentage of DNA validity (Donnelly *et al.*, 2000; Aitken *et al.*, 1992).

5-3: Prognostic factors affecting on the IUI outcome of infertile patients.

Intrauterine insemination is greatly used for the recovery of infertile patients, with wide range for success when IUI is well isolated. Sperm parameters are factors that have effects on successful the methods. The sperm motility is one of the factor for occurring higher rates of pregnancy in patients for IUI recovery (Shulman *et al.*,1998). The rates of pregnancy following IUI vary among studies which may be due to the variation sperm quality and quantity, activation methods, selection patients, different factors of infertility and ovarian stimulation techniques (Zafar *et al.*, 2007) . Also, ovarian stimulation in IUI which enhance the number of oocytes at the place of conception and enhance pregnancy rate following intrauterine insemination (Hwang *et al.,* 2003) . Moreover , sperm parameters of male partners' using in IUI generally play a predictor factors and enhance the possibility of pregnancy (Wainer *et al.*, 2004).

Many predictor factors on the pregnancy rates in intrauterine insemination have suggested , the involvement of the number of follicles, duration of infertility, thickness of endometrial, age , Ovarian stimulation, type of infertility and sperm quality (Khalil *et al.*,2001).

The results in figure (4-1) showed high value of pregnancy rate with unexplained group compared with other infertile groups. Because the unexplained infertility is one of etiology of infertility that affects the IUI outcomes , higher pregnancy rate showed with unexplained infertility, it has a best predictor in pregnancy compared with other conditions (Ahinko,2009). Therefore the result in the present study agrees with many studies indicated the IUI with unexplained infertility that showed elevation in the pregnancy rates where the rate of pregnancy was 18 % (Chung *et al.*, 1995). Another study by (Allen *et al.,(*1985) showed the pregnancy rate following IUI was 25%. Also, the pregnancy rate in IUI was significantly higher in unexplained group compared to teratozoospermia(Grigoriou *et al.*,2005). Commonly, the cause of unexplained infertility in infertile patients may be due to problems with abnormal sperm DNA, sperm capacity of fertilization, ovum quality, failure of implantation, genetic etiology and cervical factors. (Rajasheker *et al.*,2008). Therefore, the present study depends on the evaluation of different variables to improve the positive effects of the both activation methods (density gradient and swim-up) and promote the motility and to add further information on the quality of spermatozoa that will be used in future on IUI outcomes.

The major purpose of sperm activation for IUI is to isolate a lot of motile sperm with optimal form and yield sperm with best DNA and chromatin from ejaculation . Improvement above variables in final activation has been proposed to be essential of the fertilization and outcomes (Gellert – Mortimer *et al.*, 1988). The density gradient and swim – up methods have various effectiveness in isolating the sperm. Whence the sperm activation techniques (density gradient and swim – up) provide the elevation of pregnancy rate (Carrell *et al.*, 1998). The sperm quality is better activated with density gradient compared with swim – up. Elevation treatment rates of mature motile sperm occurred following sperm activation by density gradient which may be most useful for increasing pregnancy rate in assisted reproduction techniques (Shyam *et al.*, 2005).

The results in figures (4-2,4-3,4-4,4-6) found high significance in pregnancy rate with density gradient technique compared to swim-up technique which are in agreement with many studies . A number of studies were found higher pregnancy rate with density gradient centrifugation than swim – up following IUI, because density gradient has higher motility rate than swim – up, also the time used in the procedure of swim – up is longer which may decrease the number of motile sperm to accumulate. While density gradient is less time where the immotile sperms down load to the bottom of tube (Nilgun *et al.*, 2011). Moreover, sperm activation by density gradient centrifugation appeared to be the better compared to the swim – up and simple method: a best improvement and decrease DNA damage in the activated specimen (Boomsma *et al.*, 2007). A study by Ombelet(2013) indicated that used density gradient and swim – up in IUI showed a significant difference in pregnancy rate (38.1% versus 22.2% respectively).

Concerning the sperm concentration and progressive motility, the number of motile sperm, progressive motility and sperm count in both pre – activation and post – activation were significantly elevation in IUI outcome (Marshburn *et al.*, 1992). The number of motile count of spermatozoa following activation showed better correlation with normospermic outcome (Bielsa *et al.*, 1994). Our study referred to the progressive motility (A+B) following activation as a main predictor correlation with pregnancy, progression (C and D) which did not produce any pregnancy. Therefore, progressive motility (A+B) is important for fertilization capacity and pregnancy after intrauterine insemination (Hendin *et al.*, 2000).

Moreover the present data in the figures (4-9,4-10) concluded that pregnancy rate significantly increase with sperm concentration ≥ 20 million/ ml and progressive motility ≥ 50 %. This result is compatible with Aitken *et al.* (1982) who stated the best pregnancy rates show significantly higher when fresh semen specimen ≥ 20 million /ml spermatozoa, lower this concentration the correlation between fertilization capacity and sperm concentration may lead to less predictability.

Therefore, the increase of the rate of pregnancy correspond with the increase of sperm concentration and sperm motility. The percentage of inseminated total sperm concentration may be prognostic factor which a successful pregnancy rate (Pasqualoto *et al.*, 1999). A study by Zhao *et al.* (2004) indicated of achievement of higher pregnancy rate with elevation number of motile count. This study showed increased chance of IUI outcome when sperm concentration more than 20 million /ml.

In contrast, a study by Duran et al.(2002) showed no correlation between sperm parameters and IUI outcome. Also, Burr *et al.* (1996) suggested the progressive motility (A+B) and sperm count following activation have no effects on the pregnancy rate after intrauterine insemination especially when number of motile count following activation is lesser than one million /ml .

Regarding the sperm morphology, defective of sperm morphology that (teratozoospermia) is diagnosed by site, includes the head, mid pice and tail. Sperm morphology is the better factor of fertilization capacity (Speroff and Fritz, 2005). Therefore, the alteration of sperm form establishes a significantly decrease IUI outcome (Merviel *et al.*, 2010). The normal morphology percent in origin and post activated semen specimens significantly increase pregnancy rate. Also it is associated with in vitro fertilization rate (Sukcharoen *et al.*, 1995), and increase IUI outcome (Burre et al., 1996). Sperm morphology may represent a most prevalent of semen parameters with motility and sperm count (Guerif, 2010). The data in figure (4-11) showed highly significant increase in pregnancy rate with sperm morphology $\geq$ 30% . This result is compatible with a study by Wainer *et al* (2004) who reported that the sperm morphology which is less than 30% gives a pregnancy rate which may reach 5.43% compared to 18.42% of pregnancy with normal morphology.

Also, the success of pregnancy rates in IUI is elevated when the normal sperm morphology was $\geq$ 30% while quite lower when normal sperm morphology is $<$ 30% (Lee *et al.*, 2002).

In relation with Leukocyte, the impacts of increase leukocytes on sperm may lead to high levels of DNA damage by increase ROS production which may be attributed to ART particularly in infertile couples (Henkel *et al.*, 2005). Therefore, increase pregnancy rate following in vitro activation may result from improvement in semen parameters because reduce sperm DNA damage and oxidative stress by elimination of seminal leukocytes (Erenpreiss *et al.*, 2002). This agrees with Menkveld (2004) who proposed that decrease leukocytes levels and ROS levels following recovery may increase pregnancy rate.

The association between leukocytospermia and oxidative stress and between oxidative stress and DNA damage lead to decrease pregnancy rate (Henke *et al.*, 2003). There was a highly significant increase in pregnancy rate with couples have leukocyte which are less than one million/ml (figure 4-12). This in agreement with a study by Barrand – lange *et al* (2011) that reported the levels of leukocytes which are less than one million is related with increase ability of fertilization and ART outcomes, while increase leukocytes concentration may affects the capacity for sperm fertilization and reduce pregnancy rates in the infertile patients. Conversely, a study by Cottell *et al.*(2000) indicated no significant correlation between increase leukocytes and failure sperm function or pregnancy rate following assisted reproduction technology.

MDA is used as a parameter of oxidative stress and prognostic factor for expecting assisted reproduction techniques outcomes (Oral Ozay *et al.*, 2006; Hsieh *et al.*, 2006). Increase production of ROS may lead to decrease sperm function during failure of sperm metabolism, sperm motion, impairment of plasma membrane and failed initiated fertilization as well as abnormal chromatin structure and break strands of DNA (Aitken and krausz, 2001).

The result in figure (4-13) revealed significantly higher in pregnancy rate with couples have MDA concentration less than 4μmol/L compared with others have more than 4μmol/L. This result agrees with many studies were reported that higher damage of sperm DNA in the infertile couples may be due to increase oxidative stress production. This positive relationship between defect of sperm DNA and OS may lead to failed spermatogenesis which tend to produce immature spermatozoa, defective sperm chromatin and abnormal form (Sakkas *et al.*, 2002; Barroso *et al.*, 2000). Moreover, oxidative stress have been appositive correlation with sperm DNA damage but negative association with sperm quality, fertilization and development of embryo and finally effects on ART outcomes (Shen *et al.*, 2002).

DNA damage may impact on pregnancy rates in intrauterine insemination and the stability of the sperm DNA is a predictor factor for success of IUI outcome therefore, sperm DNA damage is a parameter of lower pregnancy rates. Therefore lower IUI outcomes in female inseminated with semen specimen which have higher levels of sperm DNA damage is associated with elevated rate of miscarriages and decrease pregnancy rate (Duran *et al* 2002). Similar study by Bungum *et al* (2007) that report to high levels of damaged sperm DNA is related with decrease of pregnancy rate following IUI where DNA fragmentation index (DFI) less than 30% resulting 19% of pregnancy rate compared 1.5% of pregnancy rate with DFI more than 30%.

Many studies proposed that sperm DNA/chromatin damage is associated with decrease pregnancy rates in IUI and suggested the poor possibility of achieving a spontaneous pregnancy occurring with sperm DNA damage (Giwercman et al., 2010; loft *et al.*, 2003). Moreover, studies were appeared that males where showed greatly damage of sperm DNA have decrease values of natural fecundity. Above 30% of sperm DNA damage, approximately 10% of males achieved pregnancy during one year (Evenson and Wixon,2006;

Evenson *et al.*, 2002). Also, sperm DNA damage is a correlated with decrease IVF and IUI outcomes but not with intra cytoplasmic sperm injection out comes (Zini and sigman, 2009).

In this study, the result in figure (4-14) was significantly higher in pregnancy rate with couples have sperm chromatin abnormality less than 30 % compared with others have more than 30%. This is agreement with several studies showed the importance elevation of sperm chromatin abnormality its correlation not only with pregnancy rates but with success rates of ART (Agarwal and said, 2003; carrell *et al.*, 2003; Evenson *et al.*, 2002). Higher degrees of AB staining may impact fertilization rate and pregnancy rate (Nasr – Esfahani *et al.*, 2001). Because elevation remnants of histone and protamine deficiency (Oliva, 2006).

With regard to age, the present study in figure (4-15) has found the pregnancy rate was significantly reduced with advanced men age . This results comes along with many previous studies that revealed the association between increase men age and decrease IUI outcome . Many studies indicated to relationship between advanced paternal age and pregnancy rate, 50% decline pregnancy rate with age > 35 years compared with age Less than 35 years. (Cohlen *et al.*, 2005; Mathieu *et al.*, 1995). Luna et al (2009) suggested the fertilization rate after assisted reproductive technologies were reduce significantly with men age more than 50 years old. All these observation lead to fact, advance of maternal and paternal age in modern time is important factor may affecting the successful outcome of ART (Kong *et al.*, 2012). Also Levitas *et al*(2007) demonstrated that sperm motility is a prognostic factors affecting IUI outcome of infertile patients and reach to the peak when paternal age less than 25 years old (44.4 ± 21%) while the decline motility (24.8% ± 18%) appeared with age more than 55 years. A study by Auger *et al*(1995) indicated to decrease in semen function related with advance age, the sperm concentration and motility decline (2.6% and 0.6% respectively) and 0.9% reduced in the normal morphology percent.

Advance men age is also negatively affects sperm quality (sperm count, sperm motility and morphology) and function of testicle as well as increased levels of DNA damage and chromatin defective (Jason *et al.*, 2013). The essential teratogenic impacts of chemicals and environmental pollution that may accumulate with advance age may have linked to the elevated of DNA damage with poor semen quality (Pastuszak and Lamb, 2013).

Maternal age is the most significance predictor of IUI outcome, decrease pregnancy rate related with advanced female age because the decline women fertility by decrease uterine ability and reduced the number and quality of oocyt (Mohan *et al.*, 2012). Some studies have proposed no correlation between advance age and successful of pregnancy rate (dela Rochebrochard and Thonneau, 2003; Olsen, 1990).

The duration of infertility was important predicator factor for successful rate of pregnancy. A study by Snick *et al.*,(1997) showed the 2 to 4 years of infertility which lead to reduction in rates of IUI outcome. As shown in figure (4-17), infertile patients with duration of infertility more than 6 years have significant decrease in pregnancy rate compared to patients with duration of infertility less than 6 years . This result of current study agrees with Tomlinson *et al.*(1996) who found the longer duration more than 6 years has been significantly decline with success rate of IUI outcome. Also, the infertile patients of shorter duration (less than 5 years) have higher rate of pregnancy compared with more than 5 years, (Iberico *et al.*, 2004). Other study found no significant effect on IUI outcome with advance duration of infertility (Zahra and Seddigheh, 2010).

Furthermore, as shown in figure (4-18) the pregnancy rate of non-smokers infertile patients was showed a significantly higher than those of the smokers. This finding is in agreement with many studies which showed that smoking is related with decrease semen volume, sperm concentration, progressive motility and normal morphology as well as delayed liquefaction time and increase viscosity. Also it was proposed that elevated levels of DNA damage and then decrease IUI outcome compared with non-smokers. (Alvarez, 2003; Potts *et al.*, 1999). The metabolism of smoking substances may cause inflammatory reaction in the accessory glands and then release chemical mediators may have increased leukocytes (Saleh *et al.*, 2002). Leukocytospermia may increase ROS production in seminal plasma which may decrease levels of antioxidant and induce oxidative stress (Aitken *et al.*, 1996). Other study reported that semen in smokers comprise low levels of antioxidants than that nonsmokers (Fraga *et al.*, 1996).

A study by Tulay *et al* (2013) concluded the pregnancy rate after IUI decrease significantly with men smokers because of negative effect of smoking cigarette compared with nonsmokers.

Conclusion

Conclusion

The results of this study revealed that:

1- defects in sperm parameters and sperm chromatin status as well as excessive levels of oxidative stress contribute to the etiology of infertility.

2- Semen quality after activation is the most important factor for predication of successful pregnancy in IUI.

3-The results obtained suggest that IUI with high quality sperm recovered via density gradient centrifugation and superior mode of treatment for couples with unexplained infertility.

4- The pregnancy rate following IUI is positively affected by the increase of sperm concentration, progressive motility (A+B) and normal morphology and also by decrease leukocyte concentration, MDA concentration and sperm chromatin maturity.

5- The pregnancy rate in IUI is negatively affected by increase age, duration of infertility and smoking.

6- Intrauterine insemination is a useful treatment option for many couples are non-sever male factor infertility.

Recommendations

The following suggestion are recommended for future studies:

1- These study is recommended that density gradient centrifugation is the beset activation method to isolation and improvement the quality sperm recovered of infertile men for IUI.
2- Intrauterine insemination (IUI) is recommended in couples with unexplained subfertility and non-sever male factor infertility.
3- Fertility centers should not base on the seminal fluid analysis only, but from the necessity assessment of oxidative stress and sperm DNA damage / chromatin defect which appears to be a potential tool for evaluation semen samples before commencement of IUI procedure.
4- These study is recommended that in other future studies semen levels of antioxidants to be measured of infertile patients especially whenever oxidative stress generation is involved before commencement of IUI procedure.
5- These study is recommended that age, duration of infertility and smoking with advance time contribute to the etiology of infertility. Therefore, infertile patients should attend the fertility centers following one year of infertility.
6- Studying the role and effect of hormones (FSH, LH, testosterone and prolactin) of the men on the IUI outcome.

References

- **Abasalt, H. C. ; Fatemeh, K. and Seyed, G. A. H. (2013):** Correlation of sperm parameters with semen lipid peroxidation and total antioxidants levels in astheno – and oligoastheno – teratospermic men. Iranian Red Crescent Medical Journal; 15(9): 780-5.
- **Aboulghar, M. (2005) :** The importance of fertility treatment in the developing world. BJOG . 112: 1174-6 .
- **Adham, I. M.; Nayernia, K. and Burkhardi, E. (2001):** Teratozoospermia in mice laking the transition protein 2 (Tnp2), Molecular Human Reproduction, 7 (6): 513-520.
- **Agarwal , A. ; Makker , K. and Sharma , R. (2008)** : Clinical revelance of oxidative stress in male factor infertility : an update . Am. J-Reprod . Immunol. 59(1) : 2-11 .
- **Agarwal , A. ; Varghese , AC. And Sharma , RK. (2009)** : Markers of oxidative stress and sperm chromatin integrity . Methods Mol. Biol. 590 : 377-420 .
- **Agarwal , A. and Said , T. M. (2003)** : Role of sperm chromatin abnormalities and DNA damage in male infertility . Humal Reproduction Update . 9(4) : 331-345 .
- **Agarwal, A. and Said, T. (2009)**: Sperm chromatin abnormalities and DNA damage in male infertility. Hum Reprod – Update ; 9: 331-345.
- **Agarwal, A. ; Saleh, R. A. and Bedaiwy, M. A. (2003)**: Role of reactive oxygen species in the pathophysiology of human reproduction. Fertil. Steril. ; 79: 829-843.
- **Agarwal, A. ; Sharma, R. K. ; Nallella, K. P. ; Thomas, A. J. ; Alvarez, J. G. and Sikka, S. C. (2006)** : Reactive oxygen species as an independent marker of male factor infertility . Fertil. Steril. 86: 878-885 .

- **Agarwal, A. and Allamaneni, S. S. (2011)**: Free radicals and male reproduction. J. Indian Med Assoc, 109: 184-187.
- **Agarwal, A. and Allamaneni, S. S. R. (2004)**: Oxidants and antioxidants in human fertility. Mid East Fert. Society J. , 9, 187-49.
- **Agarwal, A. and Prabakaran, S. A. (2005)**: Mechanism, measurement, and prevention of oxidative stress in male reproductive physiology. Indian. J. Exp. Biol. ; 43(11): 963-74.
- **Ahinko, K. (2009)**: Successful intrauterine insemination treatment. Finland: University of Tempere; 24 – 25.
- **Aitken , RJ.; Best , FS. and Richardson , DW. (1982)**:Ananalysis of sperm function in cases of unexplained infertility: conventional criteria , movement characteristics , and fertilizing capacity. Fertil Steril; 38:212-21.
- **Aitken, R. J. ; Baker, M. A. and Sawyer, D. (2003)**: Oxidative stress in the male germ line and its role in the a etiology of male infertility and genetic disease. Reprod Biomed Online;7:65-70.
- **Aitken, R. J. ; Deiuliis, G. N. ; Finnie, J. M. ; Hedges, A. and McLachlan, R. J. (2010)**: Analysis of the relationships between oxidative stress, DNA damage and sperm vitality in a patient population: development of diagnostic criteria. Hum Reprod, 25: 2415-2426.
- **Aitken, R. J. and Baker, M. A. (2006)**: Oxidative stress, sperm survival and fertility control. Mol. And cell Endocrin. , 250, 66-69.
- **Aitken, R. J. and Curry, B. J. (2011)**: Redox regulation of human sperm function: From the physiological control of sperm capacitation to the etiology of infertility and DNA damage in the germ line. Antioxid. Redox. Signal; 14: 367-381.
- **Aitken, R. J. and Deluliis, G. N. (2007)**: Origins and consequences of DNA damage in male germ cells. Reprod. Biomed. Online ; 14(6): 727-33.

- **Aitken, R. J. and DeIuliis, G. N. (2010)**: On the possible origins of DNA damage in human spermatozoa. Mol. Hum. Reprod;16(1):3-13.

- **Aitken, R. J. and Fisher, H. (1994)**: Reactive oxygen species generation and human spermatozoa: the balance of benefit and risk – Bioassays; 16: 259-267.
- **Aitken, R. J.; Buckingham, D. and West, K. (1992)**: Deferential contribution of leukocytes and spermatozoa to the generation of reactive oxygen species in the ejaculates of oligozoospermic patients and fertile donors. J. Reprod. Fertil. 94: 451-62.
- **Aitken, R.; Buckingham, D.; Brindle, J.; Gomez, E.; Baker, G. and Irvine, S,(1996)**: Analysis of sperm movement in relation to the oxidative stress created by leukocytes in washed sperm preparation and seminal plasma. Hum REPROD;10:2061–2071.
- **Aitken, RJ. and Krausz, C. (2001)**: Oxidative stress, DNA damage and the Y chromosome. Reproduction ; 122: 497-506.
- **AL- Inany, H. ; Azab, H. ; Elkhayat, W. ; Nada, A. ; Elkattan, E. and Abou – Setta, A. M. (2010)**: The effectiveness of clomiphene citrate in LH surge suppression in women undergoing IUI : a randomized controlled trial. Fertil. Steril.
- **Alaa, A. ; El- Ghobashy and Christopher, R. (2003)** : The Human Sperm Head . A key for Successful Fertilization . West, Journal of Andrology . Vol. 24 , No. 2 , March / April .
- **Alborzi , S. ; Motazedian , S. ; Parsanezhad , ME. And Jannati, S. (2007)**: Comparison of the effectiveness of single IUI versus double IUI percycle in infertile patients.Fertil Steril.80:595-599.
- **AL-Dujaily, S.S and Malik,K.(2013):** In vitro sperm activation for asthenospermic semen by using progesterone, pentoxifylline and Glycyrrhiza glabra extraction. Globel .J. Med. Res.3(1):1-6.

- **Aleksandra, K. ; Slawomir, K. and Ewa, B. (2004)**: Values of malondialdehyed in human seminal plasma. Progress in Med Res. , 2, 1-10.
- **Alkan, I. ; Semisek, F. ; Haklar, G. ; Kervancioglu, E. ; Ozveri, H. ; Yalcin, S. and Akdeas, A. (1997)**: Reactive oxygen species production by the spermatozoa of patients with idiopathic infertility: relationship to seminal plasma antioxidants . J. Urol 157, 140-143.
- **Allamaneni, S. S.; Agarwal, A. and Nallella, K. P. (2005)**: Charaterization of oxoidative stress status by evaluation of reactive oxygen species levels in whole semen and isolated spermatozoa. Fertil. Streil.; 83:800-803.
- **Allegra, A. ; Marino, A. ; Caffaro, F. ; Scaglione, P. ; Sammartano, F and Pizza, G. (2007)** : GnRH antagonist – induced inhibition of the premature LH suge increases pregnancy rates in IUI stimulated cycles. Aprospective randomized trial. Hum Reprod. 22: 101-108 .
- **Allen, NC.; Herbert, CM.; Maxson, WS.; Rogers, BI.; Diamond, MP and Wentz, AC. (1985):** Intrauterine insemination a critical review. Fertil steril; 44: 569 – 580.
- **Alvarez, J,G (2003):** DNA Fragmentation in human spermatozoa significance in the diagnosis and treatment of infertility. Minerva Ginecol; 55: 233 – 239.
- **American Society for Reproductive Medicine (ASRM) , (2010) :** Men's health Maryland, University of Maryland Medical Center (UMMC) press : 1-5 .
- **American Society for Reproductive Medicine (ASRM) , American Urological Association . AUA, (2001)** : A report on management of obstructive azoospermia . AUA Best Practice

Policy and ASRM Practice Committee Report . ISBN 0–9709327–0–7 (volume 3) ISBN 09649702–6–0(4 vulume set).

- **American Society for Reproduction Medicine (ASRM)(2007):** Society for Assited Reproductive Technology, 2005. Assisted reproductive technology success rate : national summary and fertility clinic reports. Atlanta, Ga: Us Department of Health and Human Services. CDC.
- **American Urological Association, Inc. (AUA), (2010)** : The optimal Evaluation of the Infertile Male . AUA Best Practice Statement . Available at URL address : http://www. auanet. org/guidelines.
- **Amorim, E. ; Torres, C. ; Graham, J.; Amorim, L. and Santos, L. (2009):** The hypoosmotic swelling test in fresh rabit spermatozoa Anim. Reprod. Sci.; 111(12): 338-343.
- **Antinori, M. ; Licata, E. ; Dani, G. ; Cerusico, F. ; Versaci, C. ; Angelo, D. and Antinori, S. (2008):** Intracytoplasmic morphologically selected sperm injection: a prospective randomized trail, Reprod. Biomed online; 16: 835-841.
- **Armstrong, J. S. ; Rajasekaran, M. ; Chamulitrat, W. ; Gatti, P. ; Hellstrom, W. J. and Sikka, S. C. (1999)**: Characterization of reactive oxygen species induced effects on human spermatozoa movement and energy metabolism . Free Radic Biol Med. , 26: 869-80.
- **Auger, J.;Kunstmann, JM.; Czyglik, F. and Jouannet, P(1995):** Decline in semen quality among fertile men in paris during the past 20 years. N. Engl J Med., 332: 281-5.
- **Aziz, N.; Agarwal, A.; Lewies – Jones, I.; Sharma, RK. and Thomas, AJ. Jr. (2004):** Novel associations between specific sperm morphological defects and leukocytospermia. Fertil. Steril., 82: 621-627.

- **Badawy , A. ; Elnashar , A. and Eltotony , M. (2009)** : Effect of sperm morphology and number on seccess of intrauterine insemination Fertil Steril. 91(3) : 777-81 .
- **Bagis, T. ; Haydrdedeoglu, B. ; Kilicdag, E. B. ; Cok, T. ; Simsek, E and Parlakgumus, A. H. (2010)**: Single versus double intrauterine insemination cycles : a randomized trial. Human Reproduction ; 25(7): 1684-1690 .
- **Baker, G. ; Liu, D. Y. and Bourne, H. (2000):** Assessment of the male and preparation of sperm for ARTs. In: Handbook of in vitro fertilization in: Trounson AO, Gardner DK, (eds) 2nd edition. CRC press, USA: 99-126.
- **Baker, M. A.; Krutskikh, A. and Aitken, R. J. (2003):** Biochemical entities involved in reactive oxygen species generation by human spermatozoa. Protoplasma. 221: 145-151.
- **Bamgbopa, K. T. (2010):** Hysteroscopy and assisted reproductive technology. Trop. J. Laparo Endoscopy. 1: 8-18 .
- **Barrand – Lange, V.; Pont, JC.; Ziyyat, A.; pocate, K.; Silfer, C.; Cedrin – Durnenn, I.; Fechtali, B.; Ducot, B and Wolf, JP. (2011):** Seminal lenkocytes are good Samaritans for spermatozoa. Fertil Steril ; 96: 1315 – 1319.
- **Barroso, G. ; Morshedi, M. and Oehninger, S. (2000):** Analysis of DNA fragmentation, plasma membrane translocation of phosphatidyl serine and oxidative stress in human spermatozoa. Hum. Reprod; 15(6): 1338-44.
- **Ben Abdallah, F. ; Dammak, I. ; Attia, H. ; Hentati, B. and Ammar – Keskes, L. (2009):** Lipid peroxidation and antioxidant enzyme activities in infertile men: correlation with semen parameter. J. Clin Lab Anal, 23: 99-104.
- **Bennetts, L. E. ; De Luliis, G. N. ; Nixon, B. ; Kime, M. ; Zelski, K. ; McVicar, C. M. ; Lewis, S. E. and Aitken, R. J. (2008):**

Impact of estrognic compounds on DNA integrity in human spermatozoa: Evidence for crosslinking and redox cycling activities. Mutat Res ; 641: 1-11.

- **Berni, A.; Meschini, R.; Filippi, S.; Palitti, F.; De Amicis, A. and Chessa, L. (2008):** L- carnitine enhances resistance to oxidative stress by DNA damage in Ataxia telangiectasia cells. Mutat. Res. 650, 165-74.
- **Bielsa, MA.; Andolz, p.; Gris, JM.; Martinez, P and Egozcue, J(1994):** Which semen parameters have apredictive value for pxegnancy in infertile conples? Hum Reprod ; 9:1887-90.
- **Bjorndahl, L.; Mortimer, D.; Barratt, CI. R.; Castilla, JA.; Menkveld,**R . and Kvist, U. (2010): Sperm Preparation Apractical Guide to Basic Laboratory Androlgy. Ist end. USA: Cambridge University press. 167-187.
- **Boomsma, C. M.; Heineman, M. J.; Cohlen, B. J. and Farquhar, C. (2007)**: Semen preparation techniques for intrauterine insemination. Systematic Reviews . 4: 1-16 .
- **Bossert, N. L. and De – Jonge, C. J. (2007)** : Sperm preparation for IVF and ICSI. In : Invitro fertilization, A. Practical Approach. David. K. Gardner. Informa Healthcare: 147-157 .
- **Bozhedomov, V. A. and Teodorovich, O. V. (2005)** : Epidemiology and causes of autoimmune male infertility.Urology.1:35-44.
- **Bradsbaw, K. D. ; Cbantilis, S. J. and Carr, B. R. (1999)** : Diagnostic evaluation and treatment algorithms for the infertile patients . In : Textbook of Reproductive Medicine . Bruce. RC, Richard EB. (Eds.) 2th Edition . Appleton and Lange, USA : 533-548 .
- **Brandes, M. ; Hamilton, C. J. ; Vander Steen, J. O. ; deBruin, J. P. ; Bots, R. S. and Nelen, W. L. (2011)** : Unexplained infertility : over all ongoing pregnancy rate and mode of conception . Hum Reprod ; 26 : 360-368 .

- **Budhan, S. P.; Katrin, N.; Katarina, J.; JoGayle, H. and David, E. W. (2006)** : The impact and potential etiology of teratospermia in the domestic cat and its wild velatives . Theriogenology . 66: 112-132 .
- **Bungum, M. (2012):** Role of sperm DNA integrity in fertility. In: Embryology – Updates and Highlights on (Lassic Topics, Pereira, L. V. (ed), In Tech. Available from: http://www. intechopenn.com/books/embryology-updates-andhighlights-on-classic-topics/role-of-sperm-dnaintegrity-in-fertility.
- **Bungum, M. ; Humaidan, P. ; Axmon, A. ; Spano, M. ; Bungum, L and Erenpresis, J. (2007):** Sperm DNA integrity assessment in prediction of assisted reproduction technology outcome, Hum Reprod. 22: 174-179.
- **Burger, R. (2013):** Evaluation of spermatozoa and tests for an assisted reproductive techniques program: Corellation with semen parameters and ART outcome. M. Sc. Thesis, University of Stellenbsch.
- **Burr, RW.; Siegberg , R.; FL aherty , SP.; Wang , XJ. And Mathews , CD(1996):** The influence of sperm morphology and the number of motile sperm inseminated on the outcome of intrauterine insemination (IUI) Statistical model for IUI success- Hum Reprod; 11:1892-6.
- **Cantieneau, A. E.; Heineman, M.J. and Cohlen, B. J. (2003)**: Single versus double intrauterine insemination in stimulated cycles for subfertile couples : asystematic review , Hum Reprod. 18(5): 941-6 .
- **Cantineau, AEP.; Cohlen, BJ. and Heinem, MJ. (2009):** Ovarian stimulation protocols (anti – oestrogens , gonadotrophins with and without GnRH agonists / antagonists) for intrauterine

insemication (IUI) in women with subfertility. Cochrane Database Syst. Rev.

- **Carrell, D. T. and Liu, L. (2001):** Altered protamine 2 expression is uncommon in donors of known fertility. But common among men with poor fertilizing capacity, and may reflect other abnormalities of spermiogenesis. J. Androl; 22: 604-10.
- **Carrell, D.T.; kuneck, P. H.; Peterson, C. M.; Hatasaka, H . H.; Jones, K.P and Campbell, B. F. (1998):** Arondomized prospective analysis five sperm preparation techniques before intrauterine insemination of husband sperm. Fertil steril; 69 (1): 122 – 126.
- **Carrell, DT.; liu, L.; Peterson, GM.; Jones, KP.; Hatasaka, HH. and Erickson, L. (2003):** Sperm DNA fragmentation is increased in couples with unexplained recurrent pregnancy Loss. Arch. Androl; 49(1): 49-55.
- **Carreras, A.; Palma, A. and Mendoza,C. (1990)**: Hypo osmotic swelling test in normo-, oligo-, astheno – and oligoasthenozoospermic before and after swim – up separation of spermatozoa. Andrologia. 22(4): 313-317.
- **Cavallini, G. (2006) :** Male idiopathic oligoasthenoteratozoospermia . Asian . J. Androl . 8(2) : 143-157 .
- **Chan, P. ; Corselli, J. ; Potton, W. ; Jacobson, J. ; Chan, S. and King, A. A. (2001):** A simple comet assay for archived sperm corrlates DNA frahmentation to reduced hyperactivation and penetration of zona free hamster oocytes. Fertil. Steril. 75: 186-192.
- **Chia, S. E. ; Ong, C. N. ; Chua, L. H. ; Ho, L. M. and Tay, S. K. (2000)** : Comparison of zinc concentrations in blood and seminal plasma and the various sperm parameters between fertile and infertile men. J. Andtol. 21: 53-57 .

- **Chung, CC.; Fleming, R and Jamieson, ME. (1995):** Randomized comparison of ovulation induction with and without intrauterine insemination in the treatment of unexplained infertility. Hum. Reprod; 10: 3139 – 3141.
- **Cocuzza, M.; Sikka, S. C.; Athayde, K. S. and Agarwal, A. (2007):** Clinical relevance of oxidative stress and sperm chromatin damage in male infertility: an evidence based analysis. Int. Broz. J. Urol. ; 33(5): 603-21.
- **Cohen – Bacrie, P. ; Belloc, S. ; Menezo, Y. J. Clement, P. ; Hamidi, J. and Benkhalifa, M. (2009)** : Correlation between DNA damage and sperm parameters : a prospective study of 1,633 patients. Fertil. Steril. 91: 1801-1805 .
- **Cohlen, B.; Cantioeau, A.; Dhoghe, T.and te Veld, E(2005)**: Multiple pregnancy after assisted reproduction. Lancet., 366: 452-3.
- **Cottell, E.; Harrison, RF and Mccaffery, M. (2000)** : Are seminal fluid microorganisms of significance or merely contaminats? Fertil Steril ; 74 : 465 – 470.
- **David, M. (2000)** : Sperm preparation methods . J. Androl. 21(3): 357-366.
- **de Kretser, D. M. ; Loveland, K. l. ; Meinhara, A. ; Simorangkir, D and Wreford, N. (1998)**: Spermatogenesis ; 13: 1-8.
- **De Lamirand, E. and Cagon, C. (1992)**: Reactive oxygen species and human spermatozoa. Depletion of adenosine triphosphate plays an important role in the inhibition of sperm motility. J. Androl. , 3: 379-86.
- **De Luliis, G. N. ; Thomson, L. K. ; Mitchell, L. A. ; Finnie, J. M. ; Koppers, A. J. and Hedges, A. (2009)**: DNA damage in human spermatozoa is highly correlated with the efficiency of chromatin remodeling and the formation of 8 – hydroxyl – 2 – deoxyguanosine, a marker of oxidative stress. Biol. Reprod ; 81: 517-524.

- **De Mateo, S. ; Gazquez, C. ; Guimera, M. ; Balasch, J. ; Meistrich, M. L and Ballesea, J. L. (2009):** Protamine 2 precursors (pre – p2), protamine 1 to protamine 2 ratio (pl/p2); and assisted reproduction outcome. Fertil. Steril ; 91(3): 715-22.
- **dela Rochebrochard, E and Thonneau, P. (2003)**: Paternal age ≥ 40 years: an important risk factor for infertility. Am J Obstet Gyecol; 189:901- 5.
- **Demirol, A. and Gurgan, T. (2007)** : Comparison of different gonadotrophin preparation in intrauterine insemination cycles for the treatment of unexplained infertility : a prospective, randomized study . Hum. Reprod. 22: 97-100.
- **Desai , N. ; Sharma , R. ; Makker , K. ; Sabaneg , E. and Agrwal , A. (2009)** : Physiologic levels of reactive oxygen species in neat semen of infertile men. Fertil. Steril. 92(5) : 1626-31 .
- **Desai, N. ; Sabanegh, E. ; Kim, T. and Agarwal, A. (2010)**: Free radical theory of aging: implications in male infertility, Urology; 75(1): 14-19.
- **Dohle, G. R. ; Diemer, T. ; Giwercman, A. ; Jungwirth, A. ; Kopa, Z. and Krauaz, C. (2010)** : Guidelines of male infertility In: European Association of Urology. April. Retrieved from: http://www.uroweb.org .
- **Dohle, G. R. ; Halley, D. J. ; Vantlemel, J. O. ; Van den Ouwel, A. M. ; Pieters, M. H and Weber, R. F. (2002)** : Genetic risk factors in infertile men with sever oligozoospemia and azoospermia . Hum Rerod (Oxford , England) ; 17: 13-16 .
- **Dohle, G. R. ; Jungwirth, A. ; Colpi, G. and Diemer, T. (2007)** : Cuide Line on male infertility . Eur. Associat. Urol. 11: 8-21 .
- **Donnelly, E. T. ; McClure, N. and Lewis, S. (1999)**: Antioxidant supplementation in vitro dose not improve human sperm motility. Fertil. Steril.; 72:4.

- **Donnelly, E. T.; O'Connel, M.; McClure, N. and Lewis, SE. (2000):** Differences in nuclear DNA fragmentation and mitochondrial integrity of semen and prepared human spermatozoa. Hum. Repro., 15: 1552-1561.
- **Drobins, E. Z. ; Zhong, C. Q. and Overstreet, J. W. (1991)** : Separation of cryopreserved human semen using sephadex columns , washing or precoll gradients. J. Androl. 12: 201-208 .
- **Du Plessis, S. S. ; Gokul, S. and Agarwal, A. (2013)**: Semen hypervisosity: causes, consquences, and cures. Center for Reproductive Medicin, Cleveland Clinic, 5: 224-231.
- **Dunson, D.B. ; Baird, D.D. and Colombo, B. (2004)** : Increased infertility with age in men and women . Obstet Gynecol. 103 : 51-6 .
- **Duran, E. H. ; Morshedi, M. ; Taylor, S. and Oehninger, S. (2002)** : Sperm DNA quality predicts intrauterine insemination outcome : A prospective cohort study. Hum Reprod ; 17(12): 3122-8 .
- **EL-Tohamy, M. (2012):** The mechanisms by which oxidative stress and Free Radical Damage Produces Male infertility. National Reseach Centre. Dokki Cairo. Egypt. Life Science Journal; 9(1): 674-688.
- **Elzanaty, S. ; Malm, J. and Giwercman, A. (2004)** : Viscoelasticity of seminal fluid in relation to the epididymal and accessory sex gland function and its impact on sperm motility . Int Andol. 27: 94-100 .
- **Erenpreiss, J. ; Spano, M. ; Erenpreisa, J. ; Bungum, M. amd Grwercman, A. (2006):** Sperm chromatin structure and male fertility:biological and clinical aspects. Asian J. Androl;8:11-29.
- **Erenpreiss, J., Hlevicka, S. and Zalkalns, J. (2002)**: Effect of leukocyte spermia on sperm DNA integrity: a negative effect in abnormal semen samples. J Androl ; 23 : 717 – 723.

- **Eskenazi, B. ; Wyrobek, AJ. ; Sloter, E.; Kidd, SA. ; Moore, L. ; Young, S. and Moore, D. (2003)** : The association of age and semen quality in healthy men. Hum Reprod, 18: 447-454 .
- **Esterhuizen , A. D. ; Franken , D. R. ; Lourens , J. C. H. ; Prinsloo , E. and VanRooyen . I. H. (2000)** : Sperm chromatin packaging as an indicator of in – vitro fertilization rates , Human Reproduction . 15(3) : 657-661 .
- **Evanson, D. and Wixon, R. (2006):** Meta – analysis of sperm DNA fragmentation using the sperm chromatin structure assay. Reprod. Biomed. Online; 12(4): 466-72.
- **Evenson, D. P. ; Jost, L. K. ; Corzett, M. and Balhorn, R. (2000)** : Characterstics of human sperm chromatin structure following an episode of influenza and high fever : a case study . J. Androl. 21: 739-746.
- Evenson, **D. P. ; Kasperson, K. and Wixon, R. L. (2007):** Analysis of sperm DNA fragmentation using flowcytometry and other techniques. Soc. Reprod. Fertil. Suppl ; 65: 93-113.
- **Evenson, DP.; Larson, KL and Jost, LK(2002):** Sperm chromatin strunture assay : its clinical use for detecting sperm DNA fragmentation in male infertility and comparisons with other techniques. J. Androl; 23(1):25-43.
- **Fauser, B. C. ; Boger, J. W. ; Hopm, W. C. and De – Hong, F. H. (1990)** : Bioactive and immunoreactive FSH in serum of normal and oligospermic men. Clin. Endocrin. , 32: 433-442 .
- **Ferlin, A. ; Raicu, F. ; Gatta, V. ; Zuccarello, D. and Palka, G. (2007)** : Male infertility role of genetic background . Reprod Biomed Online. 14: 734-45 .
- **Fernandez, J. L. ; Mouriel, L. ; Rivero, M. I. ; Goyanes, V. ; Vasquez, R. and Alvarez, J. L. (2003)**: The sperm chromatin dispersion

test: a simple method for the determination as sperm DNA fragmentation. J. Androl; 24: 59-66.

- **Filipponi , D. and Feil , R. (2009)** : Perturbation of genomic imprinting in oligozoospermia . Epigenetics . 4(1) : 27-30 .
- **Fischer, M. A. ; Willis, J. and Zini, A. (2003)**: Human sperm DNA integrity: correlation with sperm cytoplasmic droplets. Urology. 61: 206-11.
- **Fraczek, M. ; Szkutnik, D. ; Sanoka, D. and Kurpisz, M. (2001):** Peroxidation components of sperm lipid membranes in male infertility. Ginekol Pol. , 72: 73-79.
- **Fraga, CG.; Motchnik., Wyrobek, AJ. and Dmand Ames, BN. (1996):** Smoking and low antioxidant levels increase oxidative damage to sperm DNA. Mutat Res; 351:199-209.
- **Francavilla, F.; Santucci, R.; Barbonetti, A. and Francavilla, S. (2007)**: Naturally – occurring antisperm antibodies in men: interference with fertility and clinical implications . An update. Front Biosci. ; 12: 2890-2911 .
- **Francavilla, F. ; Sciarretta, F. ; Sorgentone, S. ; Necozione, S.; Santucci, R. ; Barbonetti, A. and Francavilla, S. (2009)**: Intrauterine insemination with or without mild ovarian stimulation in couples with male subfertility due to oligo/ astheno and or teratozoospermia or antisperm antibodies : a prospective cross – over trial. Frtil. Steril. Sep. ; 92(3) : 1009-11 .
- **Francavilla, S. ; D'Abrizio, P. and Rucci, N. (2000):** Fas and Fas ligand expression deranged spermatogenesis. J. Clin. Endocrinol Metab ; 85: 2692-700.
- **Franken, D. R. and Kruger, T. F. (2004) :** What is a normal spermatozoon ? In : Atlas of human sperm morphology evolution . New York ; 49-73 .

- **Garrett, C. and Baker, H. W. C. (1995) :** New fully automated system for the morphometric analysis of human sperm heads . Fertil. Steril. 63: 1306-1317 .
- **Garrido, N. ; Martinez – Conejero, J. A. ; Hauregui, J. ; Horcajadas, J. A. ; Simon, C and Remobi, J. (2009)** : Microarray analysis in sperm from fertile and infertile men without basic sperm analysis abnormalities reveals a significantly different trans criptoms . Fertil Steril ; 91(4) : 1307-1310 .
- **Gellert – Mortimer, ST.; clarke, CN.; Baker, HW.; Hyne, RV and Johnston, WI.(1988)**: Evaluation of nycodenz and percoll density gradients for the selection of motile human spermatozoa. Fertil Steril; 49: 335 – 341.
- **Geva, E.; Lessing, J. B. and Amit, A. (1998):** Free radicals, antioxidants and human spermatozoa: clinical implications. Hum Reprod; 13: 1422-1424.
- **Gholinezhad, M. and Hosseinzadeh, A. (2011):** Seminal plasma lipid peroxidation, total antioxidant capacity, and cigarette smoking in asthenoteratospermic men J. Men's Health; 8(1): 439.
- **Giwercman, A.; Lindstedt, l.; larsson, M.; Bungum, M.; Spano, M and Levine, R.J (2010)**: Sperm chromatin structure assay as an independent predictor of fertility in vivo: a case –control study. I n t J Androl; 33: 221 - 227.
- **Grigoriou, O.; Pantos, K.; Makrakis, E.; Hassiakos, D.; Konidaris, S and Creatsas, G. (2005):** Impact of isolated teratozoospermia on the outcome of intrauterine insemination. Fertil Steril ; 83: 773 – 775.
- **Guerif, F(2010):** Which assisted reproduction technigue as afunction of sperm morphology. Gynecol Obstet Fertil; 38:508-510.

- **Guzick, D. S. ; Overstreet, J. W. and Factor – Litvak, P. (2001)** : Sperm morphology, motility and concentration in fertile and infertile men, New England Journal of Medicine ; 345(19) : 1388-1393 .
- **Hamad, M. F. (2014):** Chromatin integrity in human ejaculate sperm of smokers and non – smoker patients and its relationship to seminal oxidative stress parameters. Int. J. Biosci. 4 (9): 126-41.
- **Hamada, A.; Estevesb, S.C. and Agarwal, A. (2011) :** Unexplained Male Infertility : Potential Causes and Management . Hum . Androl. , 1: 2-16 .
- **Hammadeh, M. E.; Zeginiadov, T.; Rosenbaum, P.; Georg, T.; Schmidt, W. and Strehler, E. (2001):** Predictive value of sperm chromatin condensation (aniline blue staining) assessment of male fertility. Archives of Andrology,46(2):99-104.
- **Hekmatdoost , A. ; Lakpour , N. and Sadeghi , M. R. (2009)** : Sperm chromatin integrity : etiologies and mechanisms of abnormality , assays clinical importance , preventing and repairing damage , Avicenna Journal of Medical Biotechnology , 1(3) : 147-160 .
- **Hendin, BN., Falcone, T., Hallak, J., Goldberg, J., Thomas, AJ. Jr. and Nelson, DR. (2000):** Effect of clinical and semen characteristics on efficacy of ovulatory stimulation in patients undergoing intrauterine insemination. J Assist Reprod Gent; 17:189-93.
- **Henkel, RR. and Schill , WB. (2003) :** Sperm preparation for ART. Reprod Biol Endocrinol . : 14: 108-129 .
- **Henkel, R. ; Hajimohammad, M. and Stalf, T. (2004):** Influence of deoxyribonucleic acid damage on fertilization and pregnancy . Fertil. Steril. , 81: 965-72.
- **Henkel, R. ; Kierspel, E. and Stalf, T. (2005)** : Effect of reactive oxygen species produced by spermatozoa and leukocytes on sperm

functions in non – leukocytospermic patients. Fertil. Steril. 83: 635-642.

- **Henkel., R.; Kierspel, E.; Hajmohammad, M.; stalf, T.; Hoogendijk, C.; Mehnert, C.; Menkveld, R.; Schill, W and kruger, TF. (2003):** DNA Fragmentation of spermatozoa and assisted reproduction technology. Reprod Biomed Online ; 7: 477 – 484.
- **Hinting, A.(1989).** Assessment of human sperm fertilizing ability. Ph.D. Thesis by Hinting, A.C, Rijks Universititeil. Gent, Belgium.
- **Hofmann, N. and Hilscher, B. (1990):** Use of aniline blue to assess chromatin condensation in morphologically normal spermatozoa in normal and infertile men, Human Reproduction, 6(7): 979-982.
- **Honig, SC. ; Lipshultz, L. I. and Jarow, J. (1994)** : Significant medical pathology uncovered by a comprehensive male infertility evaluation . Fertil. Steril. ; 62 : 1028-1034 .
- **Hsieh, Y. Y. ; Chang, C. C. and Lin, C. S. (2006):** Seminal malondialdehyde concentration but not glutathione peroxidase activity is negatively correlated with seminal concentration and motility. Int. J. Biol. Sci. , 2: 23-9.
- **Hughes, G. ; Murphy, M. P. and Ledgerwood, E. C. (2005):** Mitochondrial reactive oxygen species regulate the temporal activation of nuclear factor kappa B to modulate tumor necrosis factor – induced apoptosis: evidence from mitochondria – targeted antioxidants. Biochem. J. ; 389(1): 83-9.
- **Huszare, G. ; Sbracua, M. ; Vigue, L. ; Miller, D. J. and Shur, B. D. (1997):** Sperm plasma membrane remodeling during spermiogenic maturation in men: relationship among plasma membrane beta 1,4 galactosyl transferase , cytoplasmic creatine phosphokinase and creatine phosphokinase isoform rations. Biol. Repro. , 56: 1020.

- **Hwang, JL.; Hung, SC.; Chency.; Hsieh, ML.; Chen, PH and Lin, YH.(2003):** Ovarian stimulation by clomiphene citrate and hMG in combination whith cetrorelix acetrate for ICSI cycles. Hum Reprod ;18 (1):45 - 49.
- **Iberico, G.; Vique, J.; Ariza, N.; Lozano, JM.; Roca, M. And Liacer, J (2004):** Analysis of factors influencing pregnancy rate in homologous intrauterine insemination. Fertil Steril;81(5):1308-1313.
- **Irvine, D. S. (1996):** Glutathion as atreatment for male infertility. Rev. Reprod., 1: 441-447.
- **Isidori, A. ; Latini, M. and Romanelli, F. (2005)** : Treatment of male infertility. Contraception, 74 : 314-318 .
- **Jacques Auger (2010)** : Assessing human sperm morphology . Top models, underdogs or biometrics ? Asian Journal of Andrology . 12: 36-46 .
- **James, M. ; Roberts Carl, A. and Hubel (2004):** Oxidative stress in preeclampsia, 190:5.
- **Jarow, J. P. (2007) :** Diagnostic approach to the infertile male patient . Endocrinol Metab Clin North Am ; 36 : 297-311 .
- **Jarvi, K. ; Lo, K. ; Fischer, A. ; Grantmyre, J. ; Zini, A. ; Choo, V. and Mak, V. (2010)** : " CUA Guideline : The Worup of azoospermic males" . Can. Uril. Assoc. J. , 4(3): 163-167 .
- **Jason, R.; Josephine, A.; Ryan, p.; Robert, M.; Dolores, J. and Larry, I.(2013):** The effects of advanced paternal age on fertility. Asian Journal of Andro logy., 15:723 -728.
- **Johansson, M. ; Hellstrom, A. L. and Berg, M. (2011):** Sever male infertility after failed ICSI treatment – a phenomenological study of men's experiences . Reproductive Health . 8(1):4 .
- **Kaewnoonual, N.; Chiamchanya, C.; Visutakul, P. ; Mauochantr, S.; Chaiya, J. and Tor – Udom, P. (2008):** Comparative study of

semen quality between pre – washed and post – washed with 3 sperm preparation media. Thammasat Medical Journal; 8(3): 292-300.

- **Katsoff, D. ; Kiefer, D. ; Check, M. and Check, J. H. (1999)** : Oligoasthenozoospermia is not associated with spontaneous aborions following in vitro pregnancies contrasting with in vitrofertilization data. Arch. Androl. 42 – 203 – 205 .
- **Katzorke, T. and Kolodziej, F. B. (2010)** : Significance of insemination in the ear of IVF and ICSI . Urologe . 49(7): 842-846.
- **Kazerooni, T. ; Asadi, N. ; Jadid, L. ; Kazerooni, M. ; Ghanadi, A. ; Ghaffarpasand , F. ; Kazerooni, Y. and Zolghadr, J. (2009):** Evaluation of sperm's chromatin quality with acridine orange test, chromomycin A_3 and aniline blue staining in couples with unexplained recurrent abortion. J. Assist Reprod. Genet. , 26 (11): 591-596.
- **Keegan, B. R. ; Barton, S. ; Sanchez, X. ; Berkeley, A. S. ; Krey, L. C. and Grifo, J. (2007) :** Isolated teratozoospermia does not effect in vitro fertilization outcome and is not an indication for intracytoplasmic sperm injection. Fertil. Steril. ; 88: 1583-8 .
- **Khalil, MR.; Rasmussen,PE.; Erb, K.; Laursen, SB.; Rex, S. and Westergadrd, LG.(2001):** Homologous intrauterine insemination. An evaluation of prognostic factors based on a review of 2473 cycles Acta obstet Gynecol Scand;80(1):74–81.
- **Kong,A.; Frigge, ML.; Masson,G.; Besenbacher,S. and Sulem, P. (2012):** Rate of de novo mutation and the importance of fathers age to disease risk Nature; 485:246 -50.
- **Kubo – Irie, M. ; Matsumiya, K. ; Iwamoto, T. and Kaneko, S. (2004) :** Morphological abnormalities in the spermatozoa of fertile and infertile men. Mol. Reprod. 70: 70-81 .

- **Kumar, K. ; Thilagavathi, J. ; Deka, D. and Dada, R. (2012):** Unexplained early pregnancy loss: role of paternal DNA. Indian J. Med. Res. 136: 296-298.
- **Kumar, R. ; Venkatesh, S. ; Kumar, M. ; Tanwar, M. ; Shasmsi, M. B. ; Kumar, R. ; Gupta, N. P. ; Sharma, R. K. ; Talwar, P. and Dada, R. (2009)** : Oxidative stress and sperm mitochondrial DNA mutation in idiopathic Oligoasthenozoospermic men. Indian Journal of Biochemistry and Biophysics. 46(2): 172-177.
- **Kunzle, R. ; Mueller, M. D. ; Hanggi, W. ; Birkhauser, M. H. ; Drescher, H. and Bersinger, N. A. (2003)**: Semen quality of male smokers and nonsmokers in infertile couples. Fertil. Steril; 79: 287-91.
- **Lackner, J. ; Schatzl, G. ; Horvath, S. ; Kratzik, C. and Marberger, M. (2006)** : Value of counting white blood cells (WBC) in semen samples to predict the presence of bacteria. Eur. Urol. 49: 148-53.
- **Lakpour, N. ; Mahfouz, R. Z. ; Akhondi, M. M. ; Agarwal, A. ; Kharrzai, H. ; Zera'ati, H. ; Amirjannati, N. and Sadeghi, M. R. (2012):** Relationship of seminal plasma antioxidants and serum male hormones with sperm chromatin status in male factor infertility. Systems Biology in Reproduction Medicine, 58: 236-244.
- **Lambardo, F. ; Gandini, L. ; Lenzi, A. and Dondero, F. (2004) :** Antisperm immunity in assisted reproduction. J. Reprod. Immunol. , 62 : 101-109 .
- **Leduc, F. ; Nkoma, G. B. and Boissonneault, G. (2008):** Spermiogenesis and DNA repair: a possible etiology of human infertility and genetic disorders. Syst. Biol. Reprod Med. ; 54: 3-10.

- **Lee, R.; Margantis, M.; Channon, K. and Antoniades, C. (2012):** Evaluation oxidative stress considerations. Curr Med . Chem,19(16):2504-2520.
- **Lee, RK.; Hou, JW.; HO, Hy.; Hwu, YM.; Lin, MH and Tsai, YC.(2002):** Sperm morphology analysis using strict criteria as aprognostic factor in intrauterine insemination. Int J Androl; 25:277.
- **Levitas, E.; Lunenfeld, E.;Weisz, N.; Friger, M. and Potashnik, G. (2007):** Relationship between age and semen parameters in men with normal sperm concentration: analysis of 6022 semen samples. Andrologia, 3g:4s-50.
- **Lewis, S. E. and Aitken, R. J. (2005):** DNA damage to spermatozoa has impacts on fertilization and pregnancy. Cell Tissue Res. 322: 33-41.
- **Li, K. ; Shang, X. and Chen, Y. (2004):** High – performance liquid chromatographic detection of lipid peroxidation in human seminal plasma and its application to male infertility. Clin Chim Acta; 346: 199-203.
- **Liu, D. Y. (2013):** Clinical value of sperm DNA damage should be assessed in motile sperm fraction rather than whole ejaculated sperm. Pub Med. Central (United kigdom).
- **Liu, D. Y. ; Garrett, C. and Baker, H. W. G. (2003)** : Low proportions of sperm can bind to the zona pollucida of human oocytes . Hum. Reprod. 18: 2382-2389 .
- **Liu, D. Y. and Baker, H. W. (2000)** : Defective sperm – zona pellucida interaction: a major cause of failure of fertilization in clinical in– vitro fertilization . Hum Reprod ; 15: 702-708 .
- **Loft, S. ; Kold – Jensen, T. ; Hjollund, N. H. ; Giwercman, A. ; Gyllemborg, J and Ernst, E. (2003):** Oxidative DNA damage

in human sperm influences time to pregnancy Hum. Reprod ; 18: 1265-72.

- **Lorusso, F. ; Ceci, O. ; Bettocchi, S. ; Lamanna, G. ; Costantion, A. ; Serrati, G. and Depalo, R. (2008)** : Office hysteroscopy in an invitro fertization program. Gynecol. Endocrinol. 24: 462-469 .
- **Luna, M.; Finkler, E.; Barritt, J.; Bar-Chama, N. and Sandler, B.(2009):** Paternal ago and assisted reproductive technology outcome in ovum recipients. Fertil steril; 92:1772-5.
- **Macizo, S. ; Galvez, P. ; Jorquera, G. ; Peinado, R. ; Alvarez, C. ; Canteras, J. and Parrilla, P. (2012)** : Pregnancy predictors after Intranterine Insemination : Analysis of 3012 cycles in 1201 couples . J Reprod Infertil ; 13(3): 158-166 .
- **Makker, K.; Agarwal, A. and Sharma, R. (2009):** Oxidative stress and male infertility. Indian. J. Med. Res. 129: 357-367.
- **Makler, A. and Jacobi, P. (1981):** Factor affecting sperm motility. V. washing and resuspersion of human spermatozoa in various artificial media. Fertile. Steril. 35: 442-446.
- **Makler, A.; Fisher, M. ; Murillo, O.; Leufer, N.; Dechereny, A. and Naftilin, F. (1984).** Factor affecting sperm motility. IX. Survival of spermatozoa in various biological media and under different gaseous compositions. Fertile. Steril. 41:428-432.
- **Marshburn, PB.; Melntire, D.; Carr, BP. and Byrd, W.(1992):** Spermatozoal characteristics from fresh and frozen fertility outcome after intrauterine insemination. Fertile Sterile; 58:179-86.
- **Martinez – Pastor, F. ; Aisen, E. ; Fernandez – Santos, M. R. ; Esteso, M. C. ; Maroto – Morales, A. ; Garcia – Alvarez, O. and Grade, J. (2009)**: Reactive oxygen species generators affect quality parameters and apoptosis markers differently in red deer spermatozoa, Rroduction, 137: 225-235.

- **Mathieu, S.; Ecochard, R. and Bied, V. (1995):** Cumulative conception rate following intrauterine artificial insemination with husbands spermatozoa: influence of husbands age. Hum. Reprod., 10: 1090-1097:
- **Mengual, L. ; Ballesca, J. L. ; Ascaso, C. and Oliva, R. (2003):** Marked differences in protamine content and p1/p2 ratios in sperm cells from percoll fractions between patients and controls. J. Androl. ; 24: 438-447.
- **Menkveld, R. (2004):** Leukocytospermia. Int Congr Ser : 2004: 218 – 224.
- **Merviel, P.; Heraud, MH.; Grenier, N.; Lourdel, E.; Sanguinet, P. and Copin, H (2010):** Predictive factors for pregnancy afer intrauterine insemination(IUI): an analysis of 1038 cycles and a review of the Literature. Fertil Steril; 93:79-88.
- **Moedersheim, E. (2010) :** Flow Visualization of Rotor Wakes Using Density Gradient Methods. Rotorcraft Aerodynamics Group: 12-16 .
- **Moghissi, K. S. ; Sacco, A. G. and Borin, K. (1980)** : Immunologic infertility . Cervical mucus antibodies and postcoital test. Am J Obstet Gynecol; 136 : 941-950 .
- **Mohan, S.; Priya, B.; Aleyamma, TK.; Raju, N.; Chandy, A, Ann, M., Muthukumar, K. and korula, G. (2012):** Predictive factors for pregnancy after intrauterine insemination: A prospective study of factors affecting outcome. Journal of Human Reproductive Sciences., 3. 3.
- **Morrell, J. M. ; Moffatt, O. ; Sakka, S. D. ; Manicardi, G. C. ; Bizzaro, D. ; Tomlinson, M. ; Nilsson, H. ; and Holmes, P. V. (2004)** : Reduced senescence and retained nuclear DNA integrity in human spermatozoa prepared by density gradient centrifugation J Assist Reprod Genet. 21: 217-222 .

- **Mortimer, D. (2000):** Sperm preparation techniques and iatrogenic failures of in vitro fertilization. Hum. Reprod. 6: 173-176.
- **Mosammat, R. B. ; David, M. ; Salam, M. A. ; Ehsan, Q. ; Farzan, K. and Zahid, H. B. (2009)** : Antiboitic and micronutritional blend to enhance fertility potential in male having abnormal semen parameters. The open Clin. Trials. J. , 1: 7-12 .
- **Muslih, R. K. ; AL-Nimer, M. S. and AL-Zamely, O. M. (2002).** The level of malondialdehyde after activation with (H_2O_2) and ($CuSo_4$) and inhibition by desferoxamine and molsidomine in the serum of patients with acute myocardial infarction. National Journal of chemistry; 5: 139-148.
- **NAFA and ESHRE (Nordic Association for Andrology and European Society of Human Reproduction and Embryology – special Interest Group on Andrology) (2002):** Manual on Basic semen Analysis: 1-24.
- **Nasr – Esfahani, M. H.; Razavi, S. and Mardanim M.(2001):** Relation between different human sperm nuclear maturity tests and in vitro fertilization. Journal of Assisted Reproduction and Genetics; 18 (4): 219 – 225.
- **Ng, S. C. ; Bongso, A. ; Sathananthan, H. and Rathnam, S. S. (1990) :** Micromanipulation, Its relevance to human in vitro fertilization . Fertil. Steril. 53: 203-19 .
- **Nilgun, T.; Aslihan, P.; Yuksel, O.; Zehra, C. and Omer, B. (2011):** Single or double sperm wash processing by density gradient centrifngation: effect on clomiphene citrate induced intrauterine insemination cycle out comes. Turk J Med Sci ; 4(1)34 – 44.
- **Nyboe, A. and Erb, K. (2006):** Register data on Assisted Reproductive Technology (ART) in Europe including adetailed description of ART in Denmark International Journal of Andrology . 29(1): 12-16 .

- **Oehninger, S. ; Acosta, R. ; Morshedi, M. ; Philput, C. ; Swanson, R. J. and Acosta, A. A. (1990)** : Relationship between morphology and motion characteristics of human spermatozoa in semen and in the swim – up sperm fractions. J. Androl. , 11(5): 446-452 .
- **O'Flaherty, C. ; Vaisheva, F. ; Hales, B. F. ; Chan, P. and Robaire, B. (2008):** Characterization of sperm chromatin quality in testicular cancer and Hodgkin's lymphoma patients prior to chemotherapy. Hum Repord ; 23: 1044-1052.
- **Olayemi, F. O. (2010) :** A Review on some causes of male infertility . African Journal of Biotechnology, 9(20) : 2834-2842 .
- **Olesen, I. A. ; Sonne, S. B. and Hoei – Hansen, C. E. (2007)** : Environement , testicular dysgenesis and carcinoma in situ testis . Best pract. Res. Clin. Endocrinol .Metab. 21: 462-478 .
- **Oliva, R (2006):** Protamines and male infertility. Human Reproduction Update; 12 (4): 417 – 435.
- **Olsen, J (1990)** Sub fecundity according to the age of the mother and the father. Dan Med Bull; 34:281-2.
- **Ombelet,W.; Campo, R. ; Bosmans, E. and Nijs, M. (2008)** : Intrauterine insemination (IUI) as afirst – line treatment indeveloping countries and methodological aspects that might influence IUI success . Hum Reprod 23(1) 64-72 .
- **Ombelet, W. (2013)** : Evidence – based recommendations for IUI indoily practice. Middle East Fertility society Journal ; 18: 74 – 77.
- **Oral Ozay.; Kutlu Tayfun.; Aksoy Evrim.; Ficicioglu Cem.; Uslu Husamettin.; and Tugrul Smith.(2006):** The effect of oxidative stress on outcomes of assisted reproductive techingues. J. Assist Reprod Genet; 23 (2): 81 – 85.
- **Ozmen, B. ; Koutlaki, N. ; Youssry, M. ; Diedrich, K. and Al-Hasani, S. (2007):** DNA damage of human spermatozoa in assisted

reproduction: Origins, Diagnosis, Impacts and Sefety. Reprod. Biomed Online ; 14(3): 384-95.

- **Paduch, D. A. ; Bolyakov, A. and Travis, A. (2009)** : Reproduction in men with Klinefelter syndrome : the past, the present, and the future . Semin. Reprod. Med. 27: 137-148 .
- **Pasquallotto , EB.; Daitch, JA.; Handin, BN.; Falcone, T.; Thomas, AJ. Jr and Nelson, DR.(1999):** Relationship of total motile sperm count and percentage motile sperm to successful pregnancy rates following intrauterine inseminate, on.J Assist Reprod Genet; 16(9):476-82.
- **Pasqualotto, F.F.; Sharma, R.K.; Nelson, D. R.; Thomas, A. J. and Agarwal , A(2000):** Relationship between oxidative stress, semen characteristics and clinical diagnosis in men undergoing infertility investigation. Fertil. Steril;73:459-464.
- **Pastuszak, AW and lamb, DJ (2013):** Counting your sperm before they fertilize: are sperm counts really declining? Asian J Androl; 15:179 -83.
- **Pauli, E. M. ; Legro, R. S. and Demers, L. M. (2008)** : Diminished patermity and gonadal function with increasing obesity in men. Fertil. Steril. 90: 346-351 .
- **Peivandi , S. ; Karimpour , A. and Moslemizade , N. (2010)** : Effects of L- carnitine on infertile Men's spermiogream , Radomized Clinical Trial . J. Reprod . Infertil . 10(4) : 331 .
- **Perticarari, S. ; Ricci, G. ; Granzotto, M. ; Boscolo, R. ; Pzzobon, C. ; Guarnieri, S. ; Sartore, A. and Presoni, G. (2007)** : A new multiparameter flow cytometric method for human semen analysis. Hum. Reprod. 22: 485-494 .
- **Poongthai, J. ; Gopenath, T. S. and Manonayaki, S. (2009)** : Genetics of human male infertility . Singapore Med. J. , 50(4): 336-47 .

- **Potts , RJ.; Newbury, CJ.; Smith, G.; Notarianni, LJ. and Jefferies , TM . (1999):** Sperm chromation damage associated with male smoking. Mutat Res; 423:103- 111.
- **Punab, M.; Loivukene, K. and Kermes, K.(2003)**: The limit of Leucocytospermia from the microbiological viewpoint. Andrologia. 35: 271-278.
- **Rajasheker, L.; Krishna, D and patil, M. (2008):** Polycystic ovaries and infertility: our experience. J Hum Reprod Science;1(2):65–72.
- **Rosenfeld, M. E. ; Palinki, V. W. and Hetuala, Y. S. (1990):** Distribution of oxidation specific lipid protein adducts and a polipoprotein B in atherosclerotic lesions of varying severity from N/HHL rabbits. Atherosclerosis; 10: 336-49.
- **Rousseaux, J. and prevoust, R. (1995):** Molecular localization of free thiols in human sperm chromatin. Biol. Reprod ;52(5):1066-72.
- **Roux, C. ; Tripogney, C and Joanne, C.(2004)**: Sperm chromatin packaging as an indicator of in – vitro fertilization rates , Gynecologie Obstetrique et Fertilite , 32(9) : 792-798 .
- **Rumley, A. G. ; Woodward, M. ; Rumely, A. ; Rumely, J. and Lowe, G. D. O. (2004):** Plasma Lipid peroxides: relationsphips to cardiovascular risk factors and prevalent cardiovascular disease. QJ Med; 97: 809-16.
- **Said, T. M. ; Agarwal, A. ; Erenpreiss, J. ; Mahfouz, R. ; Giwereman, A. and Evenson, D. P. (2006):** Association of early apoptosis with the nuclear integrity and maturity in different seminal fractions. Cleveland , Malmo University, Malmo, Sweden.
- **Said, T. M. ; Agarwal, A. ; Sharma, R. K. ; Mascha, E. ; Sikka, S. C. and Thomas, A. Jr. (2004):** Human sperm superoxide anion generation and correlation with semen quality in patients with male infertility, Fertility and sterility, 82(4): 871-877.

- **Said, T. m. ; Agarwal, A. ; Sharma, R. K. ; Thomas, AJ. Jr. and Sikka, S. C. (2005):** Impact of sperm morphology on DNA damage caused by oxidative stress induced by beta – nicotin amide adenine dinucleotide phosphate. Fertil. Steril. ; 83: 95-103.
- **Sakkas, D. ; Manicardi, G. C. ; Tomlinson, M. ; Mandrioli, M. ; Bizzaro, D. and Bianchi, P. G. (2000)** : The use of two density gradient centrifugation techniques and the swim – up method to separate spermatozoa with chromatin and nuclear DNA anomalies. Hum. Reprod. 15(5): 1112-1116 .
- **Sakkas, D. ; Mariethoz, E. ; Manicardi, G. ; Bizzaro, D. ; Bianchi, p. and Bianchi, U. (1999)**: Origin of DNA damage in ejaculated human spermatozoa. Rev. Reprod., 4: 31-7.
- **Sakkas, D. and Alvarez, J. G. (2010):** Sperm DNA fragmentation: mechanisms of origin, impact on reproductive outcome, and analysis. Fertil. Steril. ; 93: 1027-36.
- **Sakkas, D.; Moffatt, O.; Manicardi, GC.; Mariethoz, E.,Tarozzi and Bizzaro, D. (2002):** Nature of DNA damage in ejaculated human spermatozod and the possible involvement of apoptosis. Biol Reprod; 66: 1061 – 7.
- **Saleh, R. A. ; Agarwal, A. ; Sharma, R. K. ; Said, T. M and Sikka, S. C. (2003):** Evalution of unclear DNA damage in spermatozoa from infertile men with varicocele. Fertil Steril; 80: 1431-6.
- **Saleh, R. A. and Agarwal, A. (2002):** Oxidative stress and male infertility: From research bench to clinical practice. J. Androl.,23:737-52.
- **Saleh, R.;Agarwal, A.; Sharma, R.; Nelson, D. and Tomas, A.J.(2002):** Effect of cigarette smoking on levels of seminal oxidative stress in infertile men: a prospective study. Fertile steril; 78: 491 – 499.

- **Sanchez, R.; Villagran, E.; Risopatron, J. and Celis, R. (1994):** Evaluation of nuclear damage. Asian J Androl., 13:764-8.
- **Schlegel, P. N. (2004) :** Causes of azoospermia and their management . Reprod Fertil. Dev. 16: 561-577 .
- **Schlegel, P. N. and Girardi, S. K. (1997):** In vitro fertilization for male factor infertility. J. Clin. Endocrinal. Metab. 82: 709-16.
- **Schulte, R. T. ; Phi, D. A. ; Sigman, M. and Smith, G. D. (2010):** Sperm DNA damage in male infertility: Etiology, Assays, and Outcomes. J. Assist Reprod Gent ; 27(1): 312.
- **Seli, E. ; Gardner, D. K. ; Schoolcraft, W. B. ; Moffat, O. and Sakkas, D. (2004)**: Extent of nuclear DNA damage in ejaculated spermatozoa impacts in blastocyst development after in invtro fertilization. Fertil. Steril. ; 82: 378-83.
- **Sellami, A. ; Nozha, C. ; Soumay, B. Z. ; Hanen, S. ; Sahbi, K. ; Tarek, R. and Rand Leila, K. (2013):** Assessment of chromatin Maturity in Human Spermatozoa : Useful Aniline Assay for Routine Diagnosis of Male infertility, Hindawi Publishing Corporation. Advances in Urology 8.
- **Sepaniak, S. ; Forges, T. and Monnier – Barbarino, P. (2006):** Cigarette smoking and fertility in women and men. Gyneco. Obstet. Fetil. , 34: 945-49.
- **Sharlip, ID. ; Jarow, JP. ; Belker, AM. ; Lipshutz, LI. ; Sigman, M. and Thomas, AJ.(2002)** Best practice policies for male infertility. Fertil Sertil. 77: 873-82 .
- **Sharma, R. K.; Said, T. and Agarwal, A. (2004):** Sperm DNA damage and its clinical relevance in assessing reproductive outcome. Asian. J. Androl. 6: 139-48.
- **Shazia, M. ; Muneshwar, N. and Zingade, S. (2013):** Estimation of seminal MDA levels in infertility patients. IOSR Journal of Dental Medical Sciences (IOER - HDMS) . (4) 4: 18-23.

- **Shekarriz, M.; Sharma, RK.; Thomas, A. Jr and Agarwal, A. (1995):** Positive myeloperoxidase staining (Endtz test) as an indicator of excessive reactive oxygen species formation in semen . J assist Reprod Genet, 12: 70-74.
- **Shen, HM.; Chia, SE. and Ong, CW (2002):** Evuluation of oxidative DNA damage in human sperm and its association with male infertility. J Androl; 20: 718 – 23.
- **Shilpa, H. D. and Bijoor, R. (2013):** Malondialdehyde as a marker of lipid peroxidation in acute myocardial infarction patients. MRIMS Journal of health Sciences, 1:1.
- **Shulman, A.; Hauser, R.; Lipitz, S.; Frenkel , Y.; Dor, J.; Bider, D.; Mashiach, S.;Yogev, I. and Yavetz, H.(1998):** Sperm motility is amajor determinant of pregnancy outcome following intrauterine insemination. J Assist Reprod Genet; 15: 381 – 385.
- **Shyam, S. R.; Allamaneni.; Agarwal, A.; Rama, S.; Ranganathan, P.; Rakesh, K and Sharma. (2005)**: Comparative study on density gradients and swim – up preparation techniques utilizing neat and cryopeserved spermatozoa.Asian J Androl; 7 (1): 86 – 92.
- **Siegel, MS. (1993):** The male infertility investigation and the role of the andrology laboratory. J. Reprod Med. 38: 317-334.
- **Sies, H.(1991):** Oxidative Stress: Oxidants and Anti-Oxidants., London Academic Press.
- **Sigman, M.; Lipshultz, L. and Howard, S. (2009) :** Office evaluation of the subfertile male. In : Lipshultz LI ; Howards SS ; Niederberge CS , editors . Infertility in the male. 4th ed. Cambridge, UK : Cambridge University press : 153-176 .
- **Sigman, M. and Howards, S. S. (1998)** : Male infertility. In : Walsh PC, Retick AB, Vanghan ED, eds. Cambell's urology. Seventh ed. Philadelphia : WB Saunders. 11:1299 .

- **Sikandar , R. ; Virk , S. ; Lakhani , S. ; Sahab , H. and Rizvi , J. (2005)** : Intrauterine insemination with controlled ovarian hyperst imulation in the treatment of subfertility – J Coll physicians Surg Pak. 15(12) : 782-5 .
- **Sikka, S. C. (2004)** : Role of oxidative stress and antioxidants in andrology and assisted reproductive technology . J. Androl. 25: 5-18 .
- **Singh, A. and Agarwal, A. (2011):** The role of sperm chromatin integrity and DNA damage on ,ale infertility. Open Reproductive Science Journal ; 3: 65-71.
- **Skakkebaek , N. E. ; Jorgensen , N. ; Main , K. M. ; Rajpert – DeMeyts , E. ; Leffers , H. ; Andersson , A. M. ; J , A. ; Carlsen , E. ; Mortensen , GK. ; Jensen , TK. And Toppari , J. (2006)**: Is human fecundity declining ? International Journal of Andrology . 29(1) : 2-11 .
- **Smit, M. (2011):** Clinical Application of Sperm Chromatin Structure Assessment in Andrology Patients. Ph. D. Thesis, Rotterdam University.
- **Snick, HK.; Snick, Ts.; Evers, JL and Collins, JA (1997):** The spontaneous pregnancy prognosis in untreated sub-fertile couples: the Walcheren primary care study hum reprod.12:1582 -8.
- **Souter, I. ; Baltagi, L. M. ; Kuleta, D. ; Meeker, J. D. and Petrozza, J. C. (2011)** : Women, weight, and fertility : the effect of body mass index on the outcome of superovulation /intrauterine insemination cycles . Fertil Steril ; 95(3): 1042-7 .
- **Speroff, L and Fritz, MA. (2005):** Male infertility. In : Clinical Gynecologic Endocrinology and Infertility. Baltimore, Williams and Wilkins; 30:1135-1173.
- **Stephen, EH and Chandra, A. (1998)** : Updated projections of infertility in the United States: 1995-2025 . Fertil Steril. 70(1) : 30-40 .

- **Steures, P. ; Vandersteeg, J. W. ; Hompes, P. G. ; Bossuyt, P. M. ; Habbema, J. D. and Eijkemans, M. J. (2007) :** Effectiveness of intrauterine insemination couples with an isolated cervical factor : a randomized clinical trial. Fertil Steril. ; 88: 1692-6 .
- **Sukcharoen, N.; Keith, J.; Irvine, Ds and Aitkin, RJ(1995):** Predicting the fertilizing potential of human sperm suspensions inviter importance of sperm morphology and leukocyte contamination. Fertile; 63:1293-300.
- **Swerdloff, R. E, and Wang, C. (2002)** : The testis and male sexual function. In : Goldman, L. Bennett, L. C. eds. Cecil text book of medicine . Twenty first ed. Philadelphia : WB Saunders. 1316 .
- **Talebi , A. R. ; Moein , M. R. ; Tabibnejad , N. and Gasemzadeh , J. (2008)** : Effect of varicocel on chromatin condensation and DNA integrity of ejaculated spermatozoa using cytochemical test , Andrologia . 40(4) : 245-251 .
- **Tarozzi, N. ; Bizzaro, D. ; Flamigni, C. and Borini, A. (2007):** Clinical relevance of sperm DNA damage in assisted reproduction. Reprod Biomed Online ; 14: 746-757.
- **Tasdemir, M. and Tasdemir, I. (1997)** : Outcome of intracytoplasmic injection in cases with totally morphologically abnormal spermatozoa. Proceedings of 9th World congress on Human Reproduction , Philedelphia .
- **Thomas, J.; Fishel, S.B. and Hall, J.A. (1997)**: Increased polymorphonuclear granulocytes in seminal plasma in relation to sperm morphology. Hum Reprod. 12: 2418-2421.
- **Tomlinson, M. J. ; White, A and Barratt, C. L. (1992)** : The removal of morphologically abnormal sperm forms by phagocytes appositive role for seminal leukocytes ? Hum Reprod. 7: 517-522.

- **Tomlinson, M.; Amissah – Arthur J.; Thompson, K.; Kasraie, J and Bentick, B.(1996):** Infertility: prognostic indicators for intrauterine insemination (IUI): Statistical model for IUI success. Hum Reprod; 11: 1892 – 6.
- **Tremellen, K. (2008):** Oxidative stress and male infertility – actinical prespective. Hum. Reprod. Update; 14(3): 243-58.
- **Tremellen, K. and Tunc, O. (2010):** Macrophage activity in semen is significantly correlated with sperm quality in infertile men, Int. J. Androl. 33: 823-831.
- **Tulay, I .; Petin , O.; Mehmet , I.; Semih , K.; Yildiz , U.; Begum , A.; Naeiye, E and Sezah , S (2013)** : Effect of smoking cigarette on intrauterine insemination out comes . Basic clin Sci;2:20-24.
- **Turchi , P. and Pescatori , E. S. (2005)** : Is it necessary more andrology in infertility couple 2Ed . It . J. Sex. Reprod. Med. (12): 131-3 .
- **Twigg, J. ; Irvine, D. S. ; Houston, P. ; Fulton, N. ; Michael, L. and Aitken, R. J. (1998)**: Iatrogenic DNA damage induced in human spermatozoa during sperm preparation: protective significance of seminal plasma. Mol. Hum.Reprod;4(5):439-45.
- **Uranchimeg, D. ; Akira, K. ; Yuya, Y. ; Tomoko, T. ; Yuri, Y. ; Ayako, S. ; Yu, T. ; Toshiya, M. ; Toshiynki, Y. and Minoru, I. (2011)** : Effect of semen charactevistics on pregnancy rate following intrauterine insemination. J Med. Invest.58(1):127-3.
- **Valko, M. ; Morris, H. and Cronin, M. (2005):** Metals, toxicity and oxidative stress. Curr. Med. Chem. ; 12(11): 161-208.
- **Van Voorhis, B. J. (2006)** : Outcomes from assisted reproductive technology. Obstet. Gynecol. 107: 183-200 .
- **Venkatesh, S. ; Singh, A. ; Shamsi, MB. ; Thilagavathi, J. and Kumar, R. (2011):** Clinical significance of sperm DNA damage threshold value in the assessment of male infertility. Reprod. Sci., 18: 1005-1013.

- **Verhulst, S. M. ; Cohlen, B. J. ; Hughes, E. ; Te, V. E. and Heineman, M. J. (2004)** : Intrauterine insemination for unexplained subfertility. Cochrane Database Syst. Rev. (4): CDoo 1833.
- **Villegas, J. ; Schulz, M. ; Soto, L. ; Iglesias, T. ; Miska, W. and Sanchez, R. (2005):** Influnce of reactive oxygen species produced by activated leukocytes at the level of apoptosis in mature human spermatozoa. Fertil. Steril. ; 83: 808-810.
- **Wainer, R.; Albart, M.; Dorion, A.; Bailly, M.; and Bergere, Mandlombroso, R. (2004):** Influence of the number of motile spermatozoa inseminated and of their morphology on the success of intrauterine insemination. Hum Reprod ; 19:2060-2065.
- **Wang, X. ; Sharma, R. K. ; Sikka, S. C. ; Thomas, A. J. ; Falcone, T. and Agarwal, A. (2003):** Oxidative stress is associated with increased apoptosis leading to spermatozoa DNA damage in patients with male factor infertility. Fertil. Steril. ; 80(3): 531-5.
- **Wolf, H. (1995)** : The biologic significance of white blood cells in semen . Fertil. Steril. 63: 1143-1157.
- **Wolff, H. ; Politich, J. A. ; Martinez, A. ; Haimovici, F. ; Hill. J. A. and Anderson, D. J. (1990)**: Leukocytospermia is associated with poor semen quality. Fertil. Steril., 53: 528-536.
- **World Health Organization (WHO) (1999):** WHO Laboratory Manual for the Examination of Human Semen and Sperm – Cervical Mucus Interaction. 4th ed. New York. Cambridge University Press UK: 4-59.
- **World Health Organization (WHO) (2010)** : Laboratory Manual for Examination and Processing of Human Semen (5thed). Press, Switzerland .
- **World Health Organization WHO (1992)** : Laboratory Manual for the Examination of Human Semen and Sperm – Cervical Mucus

Interaction, 3rd ed. Cambridge : Cambridge University press. 107 .

- **World Health Organization.(2003):** WHO Laboratory Manual for the Examination of Human Semen and Sperm – Cervical Mucus Interaction. Cambridge, United Kingdom : Cambridge University Press.
- **Zafar, M.; Jameel, T and Abdulloh, KW.(2007):** Impact of intrauterine insemination as first line treatment of subfertility. J. pak Med Assoc; 57 (3): 133 – 136.
- **Zahra,B. and Seddigheh,E.(2010):** Prognostic factors of pregnancy in 500 cases of intrauterine insemination in Babol, northern Iran. Royal Institute international Journal of Fertil and steril; 4(1):35-39.
- **Zalata, A. A. ; Ahmed, A. H. ; Allamaneni, S. S. R. ; Comhaire, F. H. and Agarwal, A. (2004):** Releationship between acrosin activity of human spermatozoa and oxidative stress. Asian J. Androl. , 6:313-18.
- **Zariwala, M. A. ; Knowles, M. R. and Omran, H. (2007)** : Genetic defects in ciliary structure and function. Annu Rev. Physiol. 69: 423-450 .
- **Zavos, P. M. ; Abou – Abdallah, M. ; Aslanis, P. ; Correa, J. R. and Zarmakoupis. (2000)** : Use of the Muti ZSC one – step standardized swim – up method : recovery of high quality spermatozoa for intrautrine insemination or other forms of assisted reproductive technologies. Fertil. Steril. 74: 834-835.
- **Zhao, Y.; Vlahos, N.; Wyncott, D.; Petrella, C.; Garcia, J. and Zacur, Hetal (2004)**: Impact of semen characteristics on the success of intrauterine insemination. J Assist Reprod Genet; 21:143-8.
- **Zini, A and Sigman, M. (2009):**Are tests of sperm DNA damage clinically useful? Pros and cons. J. Androl; 30(3): 219 - 29.

- **Zini, A. ; Bielecki, R. ; Phang, D. and Zenzes, M. T. (2001):** Correlations between two markers of sperm DNA integrity, DNA denat and infertile. 75: 674-677.
- **Zini, A. ; Finelli, A. ; Phang, D. and Gagnon, C. (2000):** Influence of semen prossing on human sperm DNA integrity. Urology; 56: 1081-1084.

وزارة التعليم العالي والبحث العلمي

جامـــعة الكـــوفة

كلية العلوم

تأثير معالم النطف والاجهاد التأكسدي ونضج الكروماتين النطفي في التمنية الصنعية داخل الرحم لمرضى العقم

محسن كاظم حسن المرشدي

الخلاصة:

أن الهدف من هذه الدراسة هو زيادة معدلات الحمل بعد التمنية الصنعية داخل الرحم IUI لمرضى العقم باستخدام تقنيات تنشيط كفؤة مثل تدرج الكثافات والسباحة للاعلى . وكذلك تقيم تأثير عوامل اخرى على معدلات الحمل مثل العمر ومدة العقم ونوع العقم والتدخين .

تم أجراء هذه الدراسة للمدة من كانون الثاني ٢٠١٤ الى تشرين الاول ٢٠١٤ ، اذ تضمنت الدراسة أجراء فحص السائل المنوي وفحوصات أخرى لعينة مكونة من (١٥٥) من الرجال العقيمين الذين يراجعون مركز الخصوبة في مدينة الصدر الطبية.

أظهرت النتائج وجود ارتفاع معنوي ($p<0.05$) في كل من حجم المني و تركيز النطف و الحركة التقدمية للنطف والشكل السوي للنطف في مجموعة العقم غير المفسر مقارنة بمجاميع العقم الاخرى . بينما كشفت النتائج ايضا وجود انخفاض معنوي ($p<0.05$) في كل من زمن الاماعة و تركيز كريات الدم البيض و تركيز المالون داي الديهايد وكروماتين النطف غير السوي لمجموعة العقم الغير مفسر مقارنة بمجاميع العقم الاخرى .

كما بينت النتائج وجود انخفاض معنوي ($p<0.05$) في تركيز النطف بعد التنشيط بكلا التقنيتين (السباحة للأعلى وتدرج الكثافات) مقارنة بما قبل التنشيط في كل مجاميع العقم في هذه الدراسة ، بينما لم يؤشر وجود اختلاف معنوي بين التقنيتين بعد التنشيط .

كما أظهرت النتائج حدوث تحسن معنوي ($p<0.05$) في كل من الحركة التقدمية للنطف والشكل السوي للنطف بعد التنشيط بتقنية السباحة للأعلى وتقنية تدرج الكثافات مقارنة بما قبل التنشيط في كل مجاميع العقم المدروسة ، كما تبين حدوث تحسن معنوي ($p<0.05$) في الحركة التقدمية للنطف من خلال تقنية تدرج الكثافات بشكل افضل مما في تقنية السباحة للأعلى في مجاميع العقم غير المفسر وقلة النطف ووهن النطف .

كما كشفت هذه النتائج نقصانا معنوياً ($p<0.05$) في كل من تركيز كريات الدم البيض و تركيز المالون داي الديهايد وكروماتين النطف غير السوي بعد التنشيط بتقنية السباحة للأعلى وتقنية تدرج الكثافات في كل مجاميع العقم المدروسة مقارنة بما قبل التنشيط .

ولوحظ أن هذا الانخفاض كان معنويا ($p<0.05$) وبشكل افضل من خلال تقنية تدرج الكثافات مقارنة بتقنية السباحة للأعلى في مجموعتي العقم غير المفسر وقلة النطف .

النتائج في الدراسة الحالية كذلك بينت وجود زيادة معنوية ($p<0.01$) في معدلات الحمل بعد اجراء التلقيح بالتمنية الصنعية داخل الرحم IUI في مجموعة العقم غير المفسر مقارنة بمجاميع العقم الاخرى . وايضا لوحظ تحسن معنوي ($p<0.01$) في معدلات الحمل من خلال التنشيط بتقنية تدرج الكثافات وبصورة افضل مقارنة بتقنية السباحة للأعلى .

كما أظهرت النتائج حدوث زيادة معنوية ($p<0.01$) ومعدلات عالية من الحمل في الازواج الذين يحتوي سائلهم المنوي تركيز نطف $40 \times 10^6/mL <$ وحركة تقدمية للنطف $\geq 50\ \%$ والشكل السوي للنطف $\geq 30\ \%$ بالاضافة الى تركيز كريات الدم البيض $1 \times 10^6/mL >$ وتركيز المالون داي الديهايد $4\mu mol/L >$ وكروماتين النطف غير السوي $30\% >$. وبينت نتائج هذه الدراسة حدوث انخفاض معنوي ($p<0.01$) في معدلات الحمل مع تقدم العمر $40 <$ سنة وفترة العقم $6 <$ سنة و كذلك مع الرجال المدخنين مقارنة بغير المدخنين.

استنتجت هذه الدراسة بان الخلل الحاصل في معالم المني وحالة كروماتين النطف بالاضافة الى ارتفاع مستويات الاجهاد التاكسدي يمكن ان تكون قد ساهمت في احداث حالة العقم عند المرضى في عينة الدراسة . لذلك بينت الدراسة بأن نوعية المني بعد التنشيط كانت عاملا مهماً في توقع نجاح حالات الحمل في التمنية الصنعية داخل الرحم IUI . وقد أظهرت الدراسة بأن التمنية الصنعية مع تقنية تدرج الكثافات هي وسيلة افضل لمعالجة مرضى العقم غير المفسر .